AF839693

DBP Formation Control by Modified Activated Carbons

Awwa Research Foundation

About the Awwa Research Foundation

The Awwa Research Foundation (AwwaRF) is a member-supported, international, nonprofit organization that sponsors research to enable water utilities, public health agencies, and other professionals to provide safe and affordable drinking water to consumers.

The Foundation's mission is to advance the science of water to improve the quality of life. To achieve this mission, the Foundation sponsors studies on all aspects of drinking water, including supply and resources, treatment, monitoring and analysis, distribution, management, and health effects. Funding for research is provided primarily by subscription payments from approximately 1,000 utilities, consulting firms, and manufacturers in North America and abroad. Additional funding comes from collaborative partnerships with other national and international organizations, allowing for resources to be leveraged, expertise to be shared, and broad-based knowledge to be developed and disseminated. Government funding serves as a third source of research dollars.

From its headquarters in Denver, Colorado, the Foundation's staff directs and supports the efforts of more than 800 volunteers who serve on the board of trustees and various committees. These volunteers represent many facets of the water industry, and contribute their expertise to select and monitor research studies that benefit the entire drinking water community.

The results of research are disseminated through a number of channels, including reports, the Web site, conferences, and periodicals.

For subscribers, the Foundation serves as a cooperative program in which water suppliers unite to pool their resources. By applying Foundation research findings, these water suppliers can save substantial costs and stay on the leading edge of drinking water science and technology. Since its inception, AwwaRF has supplied the water community with more than $300 million in applied research.

More information about the Foundation and how to become a subscriber is available on the Web at **www.awwarf.org.**

DBP Formation Control by Modified Activated Carbons

Prepared by:
Tanju Karanfil, **Wei Cheng**, **Yanping Guo**, **Seyed A. Dastgheib**, and **Hocheol Song**
Clemson University
Department of Environmental Engineering and Science
L.G. Rich Environmental Research Laboratory, Box 340919, Clemson, SC 29634-0919

Sponsored by:
Awwa Research Foundation
6666 West Quincy Avenue, Denver, CO 80235-3098

Published by:

Awwa Research Foundation

American Water Works Association

IWA Publishing

DISCLAIMER

This study was funded by the Awwa Research Foundation (AwwaRF). AwwaRF assumes no responsibility for the content of the research study reported in this publication or for the opinions or statements of fact expressed in the report. The mention of trade names for commercial products does not represent or imply the approval or endorsement of AwwaRF. This report is presented solely for informational purposes.

CONTENTS

TABLES

FIGURES

FOREWORD

The Awwa Research Foundation is a nonprofit corporation that is dedicated to the implementation of a research effort to help utilities respond to regulatory requirements and traditional high-priority concerns of the industry. The research agenda is developed through a process of consultation with subscribers and drinking water professionals. Under the umbrella of a Strategic Research Plan, the Research Advisory Council prioritizes the suggested projects based upon current and future needs, applicability, and past work; the recommendations are forwarded to the Board of Trustees for final selection. The foundation also sponsors research projects through the unsolicited proposal process; the Collaborative Research, Research Applications, and Tailored Collaboration programs; and various joint research efforts with organizations such as the U.S. Environmental Protection Agency, the U.S. Bureau of Reclamation, and the Association of California Water Agencies.

This publication is a result of one of these sponsored studies, and it is hoped that its findings will be applied in communities throughout the world. The following report serves not only as a means of communicating the results of the water industry's centralized research program but also as a tool to enlist the further support of the nonmember utilities and individuals.

Projects are managed closely from their inception to the final report by the foundation's staff and large cadre of volunteers who willingly contribute their time and expertise. The foundation serves a planning and management function and awards contracts to other institutions such as water utilities, universities, and engineering firms. The funding for this research effort comes primarily from the Subscription Program, through which water utilities subscribe to the research program and make an annual payment proportionate to the volume of water they deliver and consultants and manufacturers subscribe based on their annual billings. The program offers a cost-effective and fair method for funding research in the public interest.

A broad spectrum of water supply issues is addressed by the foundation's research agenda: resources, treatment and operations, distribution and storage, water quality and analysis, toxicology, economics, and management. The ultimate purpose of the coordinated effort is to assist water suppliers to provide the highest possible quality of water economically and reliably. The true benefits are realized when the results are implemented at the utility level. The foundation's trustees are leased to offer this publication as a contribution toward that end.

Walter J. Bishop
Chair, Board of Trustees
Awwa Research Foundation

Robert C. Renner, P.E.
Executive Director
Awwa Research Foundation

ACKNOWLEDGMENTS

The authors would like to express their appreciation to various parties that assisted in different aspects of the project. The technical and administrative staff of Charleston Hanahan, Myrtle Beach and Spartanburg R.B. Simms water treatment plants assisted in collection and isolation of dissolved organic matter samples used during the project.

The project advisory committee (PAC) consisted of Thomas Gillogly of Carollo Engineers, Sunil Kommineni of Macolm Pirnie, Inc., and James Kilduff of Rensselaer Polytechnic Institute. PAC members provided many valuable comments during the course of the project. My-Linh Nguyen and Djanette Khiari of AwwaRF served as project officers and facilitated successful completion of the project.

The Department of Environmental Engineering and Science at Clemson University provided the analytical facilities for the project.

EXECUTIVE SUMMARY

Controlling the formation of regulated disinfection by-products (DBPs) has been one of the major challenges for the drinking water treatment industry in the United States (US) within the past three decades. DBPs are formed as a result of reactions between naturally occurring organic matter (NOM) and oxidants/disinfectants such as chlorine. The United States Environmental Protection Agency (USEPA) has promulgated stringent regulations under a two-stage Disinfectants/Disinfection By-Products Rule (D/DBPR) to minimize the risk of DBPs formation on the public health. The Stage 2 D/DBPR that was promulgated on January 4, 2006 requires water utilities to comply with the maximum contaminant levels (MCLs) of 80 and 60 µg/L for total trihalomethanes (THM: chloroform, bromoform, dibromochloro- and dichlorobromo-methane) and the sum of five haloacetic acids (HAA_5: monochloro-, dichloro-, trichloro-, monobromo, and dibromo-acetic acid), respectively, at *each individual monitoring location* in a distribution system (i.e., locational running annual averages) in the next six years. In order to comply with current and future regulations, drinking water utilities have been exploring various strategies to minimize the DBP formation.

Two approaches are commonly used to meet the DBP regulations: (1) removal of the DBP precursors (i.e., DOM: dissolved organic matter) before any chlorine addition, and (2) use of alternative disinfectants/oxidants (e.g., ozone, chloramines, chlorine dioxide and UV light) instead of chlorine. Lower level halogenated DBPs formation by using alternative disinfectants has been reported. However, the use of alternative disinfectants may not be always feasible to apply in existing treatment plants, and they may produce other unfavorable disinfection by-products (e.g., nitrosamines) and water quality problems (e.g., nitrification and/or elevated lead levels in some distribution systems). Enhanced coagulation, granular activated carbon (GAC) adsorption and membrane filtration are the main precursor control processes available for DBP control.

Powdered or granular activated carbon adsorption has been widely used in drinking water treatment plants primarily for taste, odor and synthetic organic contaminant (SOC) removal. However, carbon adsorption has not been widely used in practice for controlling DOM due to the low equilibrium capacities and slow adsorption kinetics of activated carbons to DOM. The main reason for these drawbacks is that the majority of commercial activated carbons have been developed primarily to remove small molecular weight hydrophobic SOCs from water. As a result, many commercial carbons do not provide feasible engineering solutions for removing large molecular weight and heterogeneous mixtures such as DOM. Although many researchers have investigated DOM adsorption by GAC, only a few studies have examined systematically the factors controlling DOM adsorption. Furthermore, no systematic effort has been made to modify activated carbons for DBP control.

RESEARCH OBJECTIVES

The main objectives of this study were: (1) to conduct a systematic investigation for developing a fundamental understanding of how activated carbons should be tailored for enhanced removal of DOM from natural waters; and (2) to investigate the effectiveness of some carbon tailoring approach(es) for DBP formation control at typical drinking water treatment conditions. It is expected that a fundamental understanding of tailoring activated carbons for

DBP formation control can be used in future for production of activated carbons from various raw materials.

APPROACH

A three-phase approach was used in this project:

Phase 1: Adsorption of DOM by modified GACs

The preliminary adsorption isotherm results obtained using a series of modified and well characterized activated carbons have shown that activated carbons having *both* a net positive charge and mesoporous structures showed high degree of DOM removal from a water source in Myrtle Beach, South Carolina. It was also found that high temperature helium or ammonia treatment, and iron impregnation of a coal-based microporous and a wood-based mesoporous GAC increased their DOM uptakes from Myrtle Beach water.

The main objectives in the first phase of the study were to: (i) further evaluate these preliminary findings using a different GAC as precursor for modification and testing DOM with significantly different characteristics obtained from two water sources, (ii) examine adsorption of DOM before and after conventional water treatment processes by modified activated carbons, and (iii) characterize DOM before and after adsorption in order to gain some insight to the removal of DOM by the modified activated carbons.

Phase 2: Adsorption of DOM and DBP formation control by steam-treated GACs

Results from Phase 1 confirmed the importance of carbon porosity (specifically the surface area and pore volume in pores larger than 1 nm) in the DOM adsorption. To investigate the role of carbon porosity, steam treatment, a technique that is commonly used in activated carbon preparation, was used to modify a precursor GAC. The main objective in this phase of the study was to investigate the impact of activated carbon pore enlargement while maintaining basic surface characteristics using steam treatment on *both* DOM adsorption and DBP formation control.

A commercial activated carbon was systematically modified by high temperature steam treatment. By increasing the treatment period, it was possible to produce progressively more mesoporous activated carbons. After preliminary testing, two treatment conditions were selected to produce steam-treated GACs having distinct and more mesoporous characteristics as compared to the precursor GAC. The effectiveness of the modified carbons for both DOM removal and DBP formation control was investigated in isotherms and rapid small scale column tests (RSSCTs). In these experiments, DOM obtained from two drinking water treatment plants with low and high specific ultraviolet absorbance ($SUVA_{254}$) source waters, respectively, were used. DOM was isolated from the influent and after conventional water treatment processes (Coagulation/Flocculation/ Sedimentation (C/F/S)) using a reverse osmosis (RO) unit. The RO isolation allowed performing all isotherms and column tests with a single batch of sample collected from each water source. Therefore, the results from all these tests can be comparable with each other, without being affected from the potential changes in the nature of DOM in water sources during long term studies. The majority of RSSCTs were performed with the treated waters (i.e., after conventional water treatment processes) to obtain practically relevant information. The effluent from RSSCTs was chlorinated in order to assess the performance of GACs in terms of DBP formation control. In limited isotherm tests, a steam-treated GAC produced by another research laboratory was tested. Finally, some regeneration experiments were performed to evaluate the reuse of the steam-treated GACs for DOM removal.

Phase 3: Adsorption of DOM and DBP control by iron-impregnated GACs

One of the findings from the first phase isotherm experiments was enhanced DOM adsorption exhibited by iron-impregnated activated carbons. This was attributed to the formation of some complexes between the iron species on the carbon surface and some DOM components. In this phase, various metal loading and activation techniques, which are widely used in catalyst preparation, were tested to deposit iron species on an activated carbon surface. The main objective was to investigate the effectiveness of iron-impregnated GACs for DOM adsorption and DBP formation control using both isotherms and RSSCTs. In these experiments, the same DOM batches collected for the steam-treated GACs in Phase 2 were used. This allowed a side-by-side performance comparison between steam-treated and iron-impregnated GACs.

CONCLUSIONS

Important findings from this project are as follows:

- The results obtained in this study using both raw and treated water DOM at the typical pH range of water treatment (i.e., pH 5.5 to 8) provided direct experimental evidence that the great majority of DOM components in natural waters cannot access pores smaller than 1 nm. Therefore, when GAC adsorption is intended to use for DOM removal, GACs with large surface areas and volumes in pores >1 nm should be selected. BET surface area and total pore volume are not the important parameters in selecting GAC for DOM removal because they do not provide information about the *accessible* pore region. GAC surface chemistry also plays a role in DOM adsorption. It was shown that high surface acidity (i.e., acidic pH_{PZC} values) significantly decreases DOM adsorption. However, the effect of pore size distribution becomes more important than pH_{PZC} at $pH_{PZC}>7$.
- GACs with larger pore sizes and higher pH_{PZC} can be produced by extending steam treatment periods during carbon preparation. The steam-treated GACs showed higher DOM uptakes compared with their precursor. However, the increase in DOM uptake was not always proportional to the extent of steam treatment. The increase in DOM uptake was mainly due to the removal of DOM components with more hydrophobic character and higher $SUVA_{254}$ values.
- The physicochemical characteristics of DOM play important roles in adsorption. Enlarging carbon pores by steam treatment enhanced the DOM uptake; however extended steam treatment period and additional pore enlargement did not always result in additional DOM adsorption for all waters. Although the higher accessible pore volume and higher pH_{PZC} as a result of steam treatment resulted in higher uptake of the Myrtle Beach treated water DOM by HD-S45 than HD-S21, the difference between the Spartanburg treated water DOM adsorption by HD-S45 and HD-S21 was negligible. For both waters, HD-S21 exhibited significantly higher DOM adsorption than the virgin HD. Myrtle Beach and Spartanburg treated DOM had similar $SUVA_{254}$ and MW_w values. The increase in the DOM uptake as a

result of steam treatment was not found to correlate with simple DOM characterization parameters, such as $SUVA_{254}$ and MW_w determined through size exclusion chromatography. It appears that additional differences (e.g., the *size* and structure) in the DOM mixtures that were not fully resolved by $SUVA_{254}$ and MW_w measurements caused differences in their adsorption by activated carbons.

- Removal of high molecular weight DOM components from water with conventional treatment processes (coagulation/flocculation/sedimentation) significantly increases the operational life of GAC filters. Therefore, the impact of DOM removal during conventional treatment processes on the performance of subsequent GAC adsorption for DBP formation control should be evaluated.
- The RSSCT results for Myrtle Beach and Spartanburg water DOM at pH 6.5 and 5.5 indicated that: (1) Steam-treated carbons had a longer operation time than their precursor to reach a certain degree of breakthrough, (2) pH effect on the DOM removal depends on the characteristics of source water. For the high $SUVA_{254}$ and more hydrophobic Myrtle Beach water, decrease in pH increased the DOM uptake significantly, whereas for the low $SUVA_{254}$ and more hydrophilic Spartanburg water, the pH effect was minimal, (3) steam-treated GACs can show higher degree of THM and/or HAA_9 precursor removal. However, the extent of removal was water specific and pH dependent. In this study, for the Myrtle Beach treated water, after 18,000 bed volumes or approximately 4+ months of operation, the effluent of steam-treated HD-S45 (at pH 5.5) column had THM and HAA_9 formation well below the regulatory limits, 80 and 60 ppb respectively, and (4) UV absorbing DOM molecules were preferentially removed by the steam-treated GACs.
- From a DBP control perspective, the effectiveness of GAC adsorption should be evaluated by performing DBP formation tests. Although, from a practical point of view, batch isotherm measurement is a simple technique that can be used to screen various activated carbons for DOM removal, a dynamic RSSCT, despite its time consuming and costly experimental nature, should be performed to obtain practically relevant information about the feasibility of using activated carbon adsorption for a full-scale application. The DBP formation in the effluent of RSSCT should be tested because monitoring of TOC or UV_{254} removal alone is not going to provide conclusive information about the effectiveness of GAC adsorption for DBP control.
- Regeneration experiments showed that high temperature treatment under H_2 or N_2 atmospheres effectively restored the DOM adsorption capacity of steam-treated GACs for at least two cycles of regeneration. In contrast, the NaOH regenerated carbons showed a continuous trend of decreasing DOM uptake with increasing regeneration cycle.
- Among several iron impregnation methods, ion exchange showed the most enhanced DOM at typical water treatment conditions based on isotherm experiments. However, the impact of iron impregnation also depended on the DOM characteristics. Isotherm results showed higher DOM uptakes of both Myrtle Beach treated and raw waters by iron-impregnated carbons with the ion exchange method. However, the effect of iron impregnation was minimal for the adsorption of Spartanburg raw and treated water DOM. Since both Myrtle Beach and Spartanburg treated waters have similar $SUVA_{254}$ and MW_w values, it appears that there were some other characteristics of DOM that were responsible for the higher affinity of Myrtle Beach water DOM toward the iron-impregnated GACs.
- The advantage demonstrated by iron-impregnated carbons prepared using the ion-exchange method during batch isotherm tests was not observed in RSSCTs. The lack of iron impact

was attributed to the different adsorption patterns between the column and batch mode isotherm tests. It appears that there are some very reactive DOM components in the water that can adsorb very fast on both carbon backbone and iron sites. Since there is a continuous supply of these reactive species in the influent water of RSSCT, they can outcompete other DOM species which have higher affinity toward iron sites but probably having relatively slower sorption kinetics. On the other hand, in an isotherm bottle, there is a limited supply of reactive DOM components. Once these species adsorb, the remaining DOM species that have tendency to adsorb, especially to iron sites, can also be removed from solution, because there is enough time and no more competition for their adsorption. This explanation is also in agreement with the higher DOM adsorption capacities observed during RSSCT as compared to the isotherm experiments.

RECOMMENDATIONS TO WATER UTILITIES

Complying with the stringent DBP regulations require water utilities consider various approaches. The optimum solution will probably be system-specific. If a utility considers GAC adsorption for DBP precursor control, the following approach is recommended based on the findings in this study:

- Detailed and accurate information should be obtained about the pore size distribution and pH_{PZC} values of candidate GACs to be evaluated for DBP precursor control. Carbon manufacturers should be able to provide this information for their products. The conventional parameters such as BET surface area and total pore volume are not useful to evaluate a GAC for DOM removal. Among several candidates GACs, the carbons with large surface areas and volumes in pores >1 nm and basic pH_{PZC} values should be selected for DBP precursor control. Among several GACs with neutral or basic pH_{PZC} values, the GACs having higher pore volumes in pores > 1 nm should be preferred since the effect of surface acidity becomes less important than the pore size distribution above the neutral pH_{PZC}.
- RSSCTs combined with THM and HAA formation tests must be performed to assess and compare the performance of candidate GACs. However, batch constant-dose (isotherm) experiments can be used to evaluate the DOM uptakes of several GACs. The carbons showing the highest DOM uptake at the typical DOC concentration of the water to be treated should be selected for further RSSCTs. Performing these initial batch screening experiments at a constant dose, particle size and contact time for all GACs is important for proper comparison. During the RSSCTs, periodical monitoring of TOC and UV_{254} at the effluent, although useful, is not sufficient because it does not provide information regarding the compliance with the DBP regulations. Therefore, RSSCT effluent must be periodically collected and exposed to post-disinfection at the typical treatment plant and distribution system conditions (i.e., chlorine or chloramines, and typical residence time of the distribution system) to examine THM and HAA formation. The results of RSSCTs can be plotted as a function of operation time for a large-scale column, as done in this study, to determine the operational life of a GAC filter for THM and HAA formation control.
- Removal of high molecular weight components during conventional treatment processes significantly increases the operational time of GAC filters for DBP formation control. Therefore, the impact of conventional treatment processes on GAC adsorption and DBP formation control should be evaluated in designing and operating GAC adsorption systems.

All the tests for GAC selection and evaluation of GAC adsorption must be conducted using a typical water (e.g., after conventional treatment processes) that will enter the GAC filter during the practical application. Operating GAC filter at a low pH value (e.g., 5.5 tested in this study) can further increase the DOM removal and DBP formation control. The effects of pH on DOM removal in both conventional treatment processes and subsequent GAC adsorption need to be optimized. In all these investigations, results of THM and HAA formation tests at the simulated treatment plant and distribution system conditions must be taken as the metric to compare the effectiveness of several GACs or GAC adsorption with other alternatives for DBP formation control.

CHAPTER 1
INTRODUCTION AND BACKGROUND

INTRODUCTION AND OBJECTIVES

Controlling the formation of regulated disinfection by-products (DBPs) has been one of the major challenges for the drinking water treatment industry in the United States (US) within the past three decades. DBPs are formed as a result of reactions between natural organic matter (NOM) and oxidants/disinfectants such as chlorine. NOM is ubiquitous in fresh water supplies, and it is a heterogeneous mixture of complex organic materials including humic substances, hydrophilic acids, proteins, lipids, carboxylic acids, polysaccharides, amino acids, and hydrocarbons (Thurman, 1985; Leenheer and Croue, 2003). The dissolved components (i.e., DOM: the portion of NOM that passes through a 0.45-μm filter) constitute the most problematic fraction of NOM with regard to drinking water treatment and supply, since they are only partly removed from water by conventional treatment processes.

The United States Environmental Protection Agency (USEPA) has promulgated stringent regulations under a two-stage Disinfectants/Disinfection By-Products Rule (D/DBPR) to minimize the risk of DBPs formation on the public health (USEPA, 1998). In the Stage 1 D/DBPR, the maximum contaminant levels (MCLs) for total trihalomethanes (THM: chloroform, bromoform, dibromochloro- and dichlorobromo-methane) and the sum of five haloacetic acids (HAA_5: monochloro-, dichloro-, trichloro-, monobromo, and dibromo-acetic acid) were set at 80 and 60 μg/L, respectively, *for the running annual average of a water system* (i.e., the overall average of monitoring locations in a distribution system). It was estimated that this rule would result in as many as 140 million people receiving increased protection from DBPs (Pontius, 1999). The Stage 2 D/DBPR that was published in January 2006 requires water utilities to comply with the same MCLs at *each individual monitoring location* in a distribution system (i.e., locational running annual averages) in the next six years (USEPA, 2006). In order to comply with current and future regulations, drinking water utilities have been exploring various strategies to minimize DBP formation.

Two approaches are commonly used to meet the DBP regulations: (1) removal of the DBP precursors (i.e., DOM) before any chlorine addition, and (2) use of alternative disinfectants/oxidants (e.g., ozone, chloramines, chlorine dioxide and UV light) instead of chlorine (Glaze, 1987; Huff et al., 1965; Scheible et al., 1992; Wolfe et al., 1984). Lower levels of halogenated DBPs formation by using alternative disinfectants have been reported (Stevens et al., 1989; Symons et al., 1996; Singer et al., 1999; Spietel et al., 2004; Karanfil et al., 2006). However, the use of alternative disinfectants may not be always feasible to apply in existing treatment plants, and they may produce other unfavorable disinfection by-products (e.g., nitrosamines) and water quality problems (e.g., nitirification and/or elevated lead levels in some distribution systems).

Enhanced coagulation, granular activated carbon (GAC) adsorption and membrane filtration are the main precursor control processes available for DBP control (Jacangelo et al., 1995). Enhanced coagulation is widely used, relatively easy to implement and cost effective as compared to the other alternatives (Crozes et al., 1995; Freese et al., 2001). However, it may not solve the needs of all treatment plants for meeting the more stringent Stage 2 D/DBPR requirements, especially those having high amounts of hydrophilic DOM in the source waters.

Membrane filtration (i.e., nanofiltration and reverse osmosis) provides a solution to both DBP and some microbial water quality problems. Nevertheless, it may not be always feasible to implement in the existing water treatment plants and small water treatment systems due to the high cost and complicated plant upgrade requirements. A new type of anion exchange resin (MIEX®) has recently been developed by Orica Watercare, a South Australian Water Corporation, and the Australian Commonwealth Scientific and Industrial Research Organization specifically for the removal of DOM from drinking water (Bourke and Slunjski, 1999; Singer and Bilyk, 2002; Budd et al., 2005). It is a strongly basic, macroporous resin with an acrylic skeleton. This synthetic resin has shown a wide range of NOM removal, from <35% to >60%, in four natural waters tested in the US (Budd et al., 2005). The high cost, recovery and reuse of the resin, disposal of resin regenerant, infrastructure (e.g., resin contact and sedimentation tanks) requirements, and proprietary production of the resin by a single company constitute some of the challenges for its further implementation.

DOM adsorption by GAC is one of the recommended treatment technologies for the removal of DOM from water to comply with the requirements of D/DBPR (Jacangelo et al., 1995; Speth, 2001; USEPA, 2006). Powdered or granular activated carbon adsorption has been widely used in drinking water treatment plants primarily for taste, odor and synthetic organic contaminant (SOC) removal. However, carbon adsorption has not been widely used for controlling DOM in practice due to the low equilibrium capacities and slow adsorption kinetics of DOM by activated carbons. The main reason for these drawbacks is that the majority of commercial activated carbons have been developed primarily to remove small molecular weight hydrophobic SOCs from water. As a result most of the commercial carbons may not provide feasible engineering solutions for removing large molecular weight and heterogeneous mixtures such as DOM.

A large amount of research has been done on the methods of carbon synthesis and surface modification to prepare materials with various physicochemical properties for different types of applications. The factors that may affect the adsorption of DOM on activated carbons have been studied for years (Weber et al., 1983; Summers and Roberts, 1988a; Summers and Roberts, 1988b; Karanfil et al., 1996a; Karanfil et al., 1999; Bjelopavlic et al., 1999). However, the information in these two different types of literature (i.e., carbon synthesis and DOM adsorption) has not been systematically analyzed and integrated for developing a fundamental understanding of DOM sorption by GAC as well as strategies for tailoring activated carbons for enhanced and selective removal of DOM from natural waters. Perhaps the greatest challenge to elucidate the limits of activated carbon applications in water treatment systems is to accurately understand the impacts of "system heterogeneity" on adsorption process. The organic and inorganic characteristics of each natural water may vary substantially depending on the source of water. Moreover, the characteristics of activated carbon may vary greatly, depending on the method of preparation. Each of these variables must be individually considered in evaluating the use of activated carbon for DOM removal and DBP formation control in practical applications.

The main objectives of this study are: (1) to conduct a systematic investigation for developing a fundamental understanding of how activated carbons should be tailored for enhanced removal of DOM from natural waters; and (2) to investigate the effectiveness of some carbon tailoring approach(es) for DBP formation control at typical drinking water treatment conditions. It is expected that development of a fundamental understanding for tailoring activated carbons for DBP control can be used in the future for production of activated carbons including those from various carbonaceous waste products. Such carbons, while offering simple and cost-effective solutions to the needs of drinking water treatment industry to comply with the DBP regulations, will also represent alternatives for sustainable use of environmental resources.

BACKGROUND

Dissolved Organic Matter

DOM is ubiquitous in surface and ground waters. It is a heterogeneous mixture of complex organic materials including humic substances, hydrophilic acids, proteins, lipids, carboxylic acids, polysaccharides, amino acids, carbonhydrates and hydrocarbons that have attached amide, carboxyl, hydroxyl, ketone, and various minor functional groups (Aiken, 1985; Malcolm, 1985; Thurman, 1985; Glaze, 1990; MacCarthy et al., 1990; Leenheer and Croue, 2003). In aquatic ecosystems, the sources of DOM can be either autochthonous (i.e., derived from biota growing within the water body such as algae and bacteria) or allochthonous (i.e., entering the system from terrestrial watershed). The concentration, composition and physiochemical characteristics of DOM are source specific. They are influenced by several biogeochemical processes occurring in aquatic environments including autochthonous carbon fixation by algae and aquatic plants, transformation and biodegradation by heterotrophic microbial activity, photodegradation by incident UV light, and partitioning between solid and liquid phases (Aiken and Cotsaris, 1995). In addition, runoff and storm events flushing organic matter from soil and plant debris, and living or decayed vegetation also contribute to the organic matter matrix in natural waters (Krasner et al., 1996). The DOM components derived from soil humus and terrestrial plants usually have relatively large amounts of aromatic carbon, whereas microbially derived DOM components have lower degree of aromaticity.

DOM can be divided into humic and non-humic fractions. The humic fraction is more hydrophobic and consists of humic acids (insoluble at pH < 2) and fulvic acids (soluble at entire pH range) having carboxylic and phenolic moieties. The non-humic fraction is more hydrophilic in character and includes biochemically well-defined compounds such as polysaccharides, proteins, peptides, amino acids, hydrophilic acids, lower molecular weight acids, lipids, and carbonhydrates. The humic fraction has been shown to be more amenable to removal by coagulation or GAC adsorption than the non-humic fraction (Owen et al., 1995). The non-humic fraction may be responsible for a greater proportion of biodegradable organic carbon, which supports bacterial regrowth in the distribution system.

The molecular conformational structure of DOM is commonly believed to be coiled and long-chain polymers with aromatic rings, which are slightly cross-linked (Stevenson, 1994). However, Conte and Piccolo (1999) also concluded that dissolved humic substances appear to be randomly self-associating small heterogeneous molecules held together through weak hydrophobic forces, rather than being polymeric coils. Therefore, there is still debate about the exact configurations of humic substances or DOM. Acidic functional groups, such as carboxylic or phenolic type functionalities, are attached to the DOM rings and can ionize as a function of pH. The absence of discrete and identifiable inflection points during potentiometric acid/base titrations of humic substances provides evidence for the heterogeneous and multicomponent properties of DOM. Values from 3 to 11 have been reported for the acid ionization constant (pK_a) of DOM functional groups (Thurman, 1985). DOM molecules carry a net negative charge around the neutral pH values of typical water treatment conditions. The negatively charged functional groups are responsible for the mutual repulsion and expansion of the coiled structure (Cameron et al., 1972; Ghosh and Schnitzer, 1980). From the polymeric structure point of view, at high concentration, low pH, or high ionic strength, DOM molecules are likely to coil densely, while they behave like flexible polymers at low concentration, neutral pH and low ionic strength.

There are several traditional parameters currently used to characterize DOM in natural waters. Dissolved organic carbon (TOC) is the analytical technique used to quantify the amount of DOM present in the water. The mean TOC concentration in surface and ground waters used for drinking water treatment in the United States has been reported to be 2.8 mg/L, with minimum and maximum concentrations being <0.35 and 27.5 mg/L, respectively (McGuire et al., 2002). The particulate organic carbon (i.e., the portion that does not pass 0.45μm filters) usually accounts for a minor fraction (<15 %) of the total organic carbon (Thurman, 1985; Karanfil et al., 2005). Specific UV absorbance at 254 nm ($SUVA_{254}$) is defined as the UV_{254} absorbance divided by the TOC concentration of the sample and reported in units of L/mg-m. By combining both UV_{254} and TOC into a single parameter, $SUVA_{254}$ is used as an indicator of the aromatic character of the organic matter. A high $SUVA_{254}$ value indicates a hydrophobic nature of DOM in a sample (Edzwald and van Benschoten, 1990; White et al., 1997; Kitis et al., 2002). Water samples with high $SUVA_{254}$ values (>3 L/mg-m) contain more hydrophobic and aromatic DOM components than low $SUVA_{254}$ waters (<3 L/mg-m) that are, in general, rich in hydrophilic DOM components. DOM can be fractionated into hydrophobic and hydrophilic components using XAD-type non-ionic macroporous resins (Thurman and Malcolm, 1981; Amy et al., 1987; Leenheer and Croue, 2003). The hydrophobic fractions of DOM, mostly in humic and highly aromatic character, are strongly adsorbed on the XAD-8 resin, while hydrophilic fractions are not retained. Hydrophobic fractions (i.e., isolated by XAD-8 resin) have been found to significantly contribute to DBP formation (Andrews, 1993; Sinha et al., 1997; Korshin et al., 1997; Kitis et al., 2002; Liang and Singer, 2003). However, hydrophilic DOM components (i.e., those not retained on XAD-8 resin) have also been shown to produce appreciable amount of DBPs in some waters (Andrews, 1993; Sinha et al., 1997; Kitis et al., 2002).

Apparent molecular weight (MW) distribution is another parameter commonly used for characterizing DOM (Amy et al., 1992). The MW distribution of aquatic DOM may range from a few hundred to 100,000 daltons (Da) (Leenheer and Croue, 2003). High-pressure size exclusion chromatography (HPSEC) or sequential membrane ultrafiltration are commonly used to determine the apparent MW distribution of DOM (Chin et al., 1994; Kilduff et al., 1996; Pelekani et al., 1999; Schafer et al., 2002). HPSEC is a fractionation method based on the diffusion of particles in a porous media packed in a column. Since large size molecules are excluded from the internal pores of the column media, they have short retention times, and thus elute from the column earlier than small size molecules. An UV or TOC detector is connected at the end of the column to quantify the DOM fractions eluting from the column. A wavelength range of 220~280 nm was recommended for the determination of DOM by UV detector (O'Loughlin and Chin, 2001). HPSEC does not yield the absolute size or molecular weight distribution of DOM. It requires the use of calibration standards that are expected to mimic the fractionation patterns of DOM molecules in the column. Another limitation of the technique is using the UV detector since there may be some fractions in the DOM mixture that do not absorb UV light. The MW distribution of DOM can be used to assess its removal in water treatment processes. Conventional water treatment processes preferentially remove high molecular weight and more hydrophobic DOM components from water (Randtke, 1988; Owen et al., 1995; White et al., 1997; Vuorio et al., 1998; Letterman et al., 1999; Matilainen et al., 2002). Although it is not possible to measure an absolute size distribution for DOM, the average sizes of DOM components have been predicted to be between 0.5 and 5 nm in various studies using different techniques (Cameron et al., 1972; Thurman et al., 1982; Aiken and Malcolm, 1987; Reckhow et al., 1990; Karanfil et al., 1999). Moore et al. (2001) compiled the sizes of aquatic DOM reported by several researchers and reported a range from 5.5 to 63 Å.

The chemical composition of DOM molecules can be characterized using a range of proximate (e.g. elemental analysis) and sophisticated spectroscopic techniques including pyrolysis GC/MS, ^{13}C-NMR and IR/FTIR (Leenheer and Croue, 2003). Pyrolysis GC/MS yields information about structural building blocks (e.g. proteins, polyhydroxy aromatics, polysaccharides, and aminosugars) while ^{13}C-NMR and IR/FTIR can provide data about aromatic content, carboxylic content and functional groups.

Granular Activated Carbon

GAC is a heterogeneous and porous adsorbent having high capacity for removing various small molecular weight organic compounds from water. Activated carbon adsorption was initially employed for taste and odor control in drinking water treatment in the late 1920s (Baylis, 1929). Since the implementation of the Safe Drinking Water Act, it has been commonly used to remove organic compounds of health concern, in addition to taste and odor control (Culp and Clark, 1983). GAC can be produced from a wide range of carbonaceous raw materials, including sawdust, peat, lignite, coal, cellulose residues, hardwoods, petroleum coke, spent ion exchange resins, sewage sludge, old automobile tires, grain hulls, nut shells and corn cobs (Lambiotte, 1942; Whittaker and Grindstaff, 1974; Sanga, 1975; Young, 1996; von Blucher and de Ruiter, 1999; Teng and Wang, 2000). The raw materials may be activated by either a stream of activating gas (e.g., steam or carbon dioxide) or a chemical agent (e.g., phosphoric acid) to produce highly porous activated carbons (Krczil, 1937; Dubinin and Zaverina, 1949; Hansen et al., 1953; Suzuki, 1954). While the pore structure is determined by the raw materials, the type of activating agent and the activation temperature and time have major influences on the activated carbon properties (Hashimoto et al., 1979).

The structure of activated carbon is commonly considered to be comprised of randomly orientated crystallites which are extensively interconnected by tetragonally bonded carbon cross-links (Smisek and Cerny, 1970). The crystallites are parallel layers of trigonally covalent bonded carbon. The distance and voids between the crystallites comprise the micropores. The ability of activated carbon to adsorb large quantities of material is directly related to its porous structure and extraordinarily large surface area (Summers, 1988b; Baker et al., 1992). The physical characteristics of activated carbon are commonly described with surface area, pore size, pore volume and their distribution.

According to the International Union of Pure and Applied Chemistry (IUPAC) classification of pore diameter, carbon pore size is classified as micropores (<20Å), mesopores (20Å~500Å) and macropores (>500Å). The micropores may be further divided as primary micropores (<10Å) and secondary micropores (10-20Å). Micropores contribute mostly to the large specific surface area of GAC and adsorption of small size molecules (e.g., SOCs). Macropores have less specific surface area, thus contribute very little to adsorption, but they are important for the access of the molecules to micropore regions. From an adsorption free energy point of view, compounds are preferentially adsorbed in pores with sizes similar to their dimensions because there are more contact points between the compound and carbon surface (Karanfil and Kilduff, 1999; Pelekani and Snoeyink, 1999).

The chemical nature of activated carbon is affected by its precursor and activation process. Generally, 5-20% by weight of activated carbon consists of non-carbon elements, with metal and surface bound oxygen as the two major components (Sontheimer et al., 1988). The surface oxygen-containing functional groups occur at the crosslinks and edges of crystallites (Puri, 1980). The edge atoms of the crystallite were approximately twenty times more reactive than those of the basal plane (Henning, 1962). Some researchers have identified the existence of carboxylic, hydroxyl, carbonyl, lactone and quinone groups on the activated carbon surface (Garten and Weiss, 1957; Boehm, 1974; Puri, 1970; Coughlin and Ezra, 1968). Sontheimer et al. (1988) summarized the possible surface oxygen-containing functional groups as shown in (Figure 1.1). Some of these oxygen-containing acidic surface functional groups can be quantified by the Boehm acid/base neutralization technique (Boehm, 1974). Most of these functionalities contribute to the surface acidity of the carbon. The surface basicity of activated carbons can be associated with resonating π electrons of carbon aromatic rings that attract protons, and basic surface functionalities (e.g., amine groups) that are capable of binding with protons. Basic oxygen-containing groups such as chromene- and pyrone-type functionalities can also contribute to the basicity of activated carbon, but their contribution is less significant in comparison to those of the π electrons of graphene layers (Mendendez et al., 1996a).

Adsorption of DOM by GAC

The adsorption mechanisms of DOM on GACs are complex and depend on various factors (Lee et al., 1981; Weber et al., 1983; Summers and Roberts, 1988a and 1988b; Karanfil et al., 1996a and 1996b; Newcombe et al., 1997a; Newcombe et al., 1997b; Newcombe and Drikas, 1997; Karanfil et al., 1998; Bjelopavlic et al., 1999; Karanfil, 2006). These factors will be discussed under two categories of interactions: physical and chemical interactions. In addition, the role of a number of factors on DOM adsorption is summarized in Table 1.1.

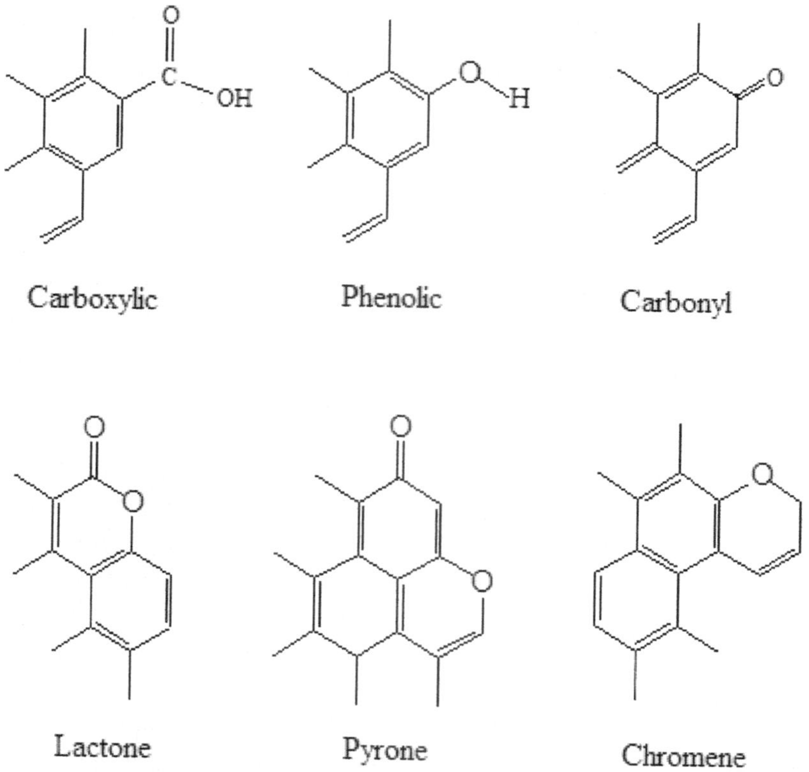

Figure 1.1 Oxygen-containing functional groups associated with activated carbon surface.

Physical Interactions

From a physical interaction perspective, the relative size distributions of activated carbon pores and DOM components are important for adsorption. For example, Filtrasorb 400 (F400), one of the most commonly used GAC in the water treatment industry, has about 1000 m^2/g specific surface area. Approximately 75% of this area is located within pores smaller than 20 Å in width (Karanfil and Kilduff, 1999). On the other hand, as mentioned before, the average sizes of aquatic DOM components have been reported to range between 5 and 60 Å in various studies (Cameron et al., 1972; Thurman et al., 1982; Aiken and Malcolm, 1987; Reckhow et al., 1990; Karanfil et al., 1999b; Moore et al., 2001). It is evident that some fractions of DOM will be excluded from adsorption. Pelekani and Snoeyink (1999) demonstrated that DOM from a ground water was excluded from primary micropores (i.e., <10 Å), but had access to secondary micropores (i.e., 10-20 Å), by studying the DOM adsorption on two activated carbon fibers with different average pore widths. It has also been shown that total accessible surface area and pore volume for DOM adsorption is inversely proportional to macromolecular size (McCreary and Snoeyink, 1980; Lee et al., 1981; Summers and Robert, 1988a and 1988b; Cornel et al., 1986; Kilduff et al., 1996). It is also important to note that large molecular weight components of DOM are preferentially removed from water during conventional water treatment processes. Although the size range of residual DOM components after such processes is not exactly known, they are still expected to be several times larger than various small molecular weight organic contaminants. Overall, considering the complex, heterogeneous and macromolecular nature of DOM components, both adsorption kinetics and equilibrium capacities of such molecules will benefit from activated carbons with pore sizes large enough to facilitate access to and the use of a greater fraction of the carbon surface area. Therefore, activated carbons with pore widths > 10 Å should be developed and used for enhanced DOM adsorption from natural waters.

Chemical Interactions

From chemical interactions perspective, the molecular structure of DOM, surface chemistry of activated carbon and solution chemistry are important variables affecting the adsorption. The majority of the oxygen-containing functionalities (e.g., carboxylic and phenolic) on the carbon surface are acidic, while basicity of activated carbons is attributed to the π electrons of carbon graphene layers or basic functionalities (e.g., basic nitrogen-containing groups). Surface functional groups influence adsorption properties and reactivity of activated carbons (Karanfil, 2006). Similar to activated carbons, various functionalities including carboxylic- and phenolic-type also exist in the DOM structures (Thurman, 1985). Higher content of these types of functionalities make the DOM components more hydrophilic. Conventional treatment processes can remove a large portion of the hydrophobic DOM components. It has been shown that the residual DOM concentration after coagulation is proportional to the hydrophilic DOM content of the water (Edzwald and van Benschoten, 1990; Owen et al., 1993; White et al., 1997). Therefore, under typical water treatment conditions, similar types of functional groups on the GAC surface and in the DOM structure play important roles in adsorption. Repulsive forces between like-charged functional groups on the activated carbon surface and DOM molecules prevent the approach of DOM to activated carbon surface, thus reducing the amount of adsorption, given that the pore size exclusion is not a factor. It has been shown that surface oxidation decreased DOM adsorption (Summers and Roberts, 1988b; Karanfil et al., 1999; Dastgheib et al., 2004), and subsequent removal of surface acidic groups by heat treatment restored the capacity (Karanfil et al., 1999; Dastgheib et al., 2004). Therefore, overcoming these repulsive interactions is important to increase the extent of DOM adsorption. Moreover, the lateral electrostatic repulsion between adsorbed DOM molecules, and repulsive interactions between adsorbed DOM and DOM in solution may also limit their adsorption.

A number of studies have revealed significantly enhanced uptake of humic substances and other DOM by GAC with increasing ionic strength (McCreary and Snoeyink, 1980; Weber et al., 1980; Summers and Roberts, 1988b; Karanfil et al., 1998). Lafrance and Mazet (1989) studied the impact of sodium salts on the adsorption of humic substances and concluded that the sodium ion concentration was the primary factor responsible for the increased adsorption of humic substances on GAC, probably via the neutralization of negative charges on both humate molecules and carbon surface functional groups. The anion of the salts had little effect on adsorption capacity. Newcombe and Drikas (1997) studied the DOM adsorption on two activated carbons at high and low ionic strength over a range of pH values. They found that at high DOM concentrations, an increase of ionic strength increased the adsorption at all pH values. The increase was more pronounced at high pH as the electrostatic repulsions between DOM molecules were higher in that pH region and were effectively screened by added salt. It was found that divalent cations, particularly Ca^{2+} and Mg^{2+}, exerted greater influence than monovalent ions (Weber et al., 1980; Kilduff et al., 1996). Karanfil et al. (1998) showed greater DOM sorption in the presence of calcium than sodium at the same ionic strength. These results suggest that the presence of Ca^{2+} may contribute more to the aggregation and coiling of humic substances in the aqueous phase, and may be effective at screening the charge of DOM molecules.

Enhanced adsorption of DOM by GAC with decreasing pH has also been reported by many researchers (McCreary and Snoeyink, 1980; Lee et al., 1981; Randtke and Jepsen, 1982; Semmens et al., 1986; Karanfil et al., 1998). This was attributed to a combination of the coiling of DOM molecules, increase in the hydrophobicity of DOM, and increasing positive charge on the carbon surface with decreasing pH. The major factor responsible for the increase of DOM uptake with increasing ionic strength and decreasing pH was thought to be the decrease in the effective molecular size of the DOM molecules, which led to greater accessibility of these macromolecules to surfaces within the carbon pores (Kilduff et al., 1996). Some researchers studied the role of dissolved oxygen (DO) on the adsorption of DOM by GAC. It was reported that the presence of DO could significantly enhance the sorption of some DOM (Warta et al., 1995; Karanfil et al., 1994; Karanfil et al., 1996b; Karanfil et al., 1998). This was attributed to the oligomerization of DOM on GAC surfaces in the presence of oxygen.

Table 1.1

The role of various factors on DOM adsorption by GAC

Factors		The Impact on DOM Adsorption
Water Chemistry	pH	**Decrease in pH increases DOM adsorption** *(A combination of the coiling of DOM molecules, increase in the hydrophobicity of DOM, and increasing positive charge on the carbon surface with decreasing pH)*
	Ionic Strength	**Increase in ionic strength increases DOM adsorption** *(A combination of screening of electrostatic repulsions between DOM molecules and GAC surface, adsorbed DOM and unadsorbed DOM molecules, and coiling of DOM)*
	Monovalent vs. Divalent Cations	**Divalent cations show higher degree of DOM adsorption than monovalent cations at the same ionic strength** *(Divalent cations are more effective than monovalent cations in the aggregation and coiling of humic substances in the aqueous phase)*
	Oxygen	**Presence of dissolved oxygen enhances adsorption of some DOM.** *(Oligomerization of DOM molecules on the GAC surfaces may occur)*
DOM	Size (or Molecular Weight) Distribution	**Smaller size DOM molecules have greater access to deeper region of GAC pores and may exhibit higher degree of adsorption**
	Chemistry	**The DOM adsorption by GAC decreases with increasing intensity of acidic functional groups in the DOM structure** *(The repulsive electrostatic forces between the same charge groups on the GAC surface and DOM decreases adsorption. Hydrophilic DOM are also expected to have higher solubility in aquatic solution)*
GAC	Surface Area	**Increase in the *Accessible* surface area increases DOM adsorption**
	Pore Size Distribution	**DOM adsorption is expected to increase with increasing pore volumes in pores > 1nm. However, pore size distribution in the pores >1 nm with respect to DOM size distribution further influence adsorption of DOM components**
	Surface Chemistry	**Increase in pH_{PZC} (or decrease in surface acidity) increases DOM adsorption** *(As pH_{PZC} increases, the net positive charge on the GAC surface increases, enhancing the adsorption of negatively-charged DOM at typical pH value of water treatment)*

Activated Carbon Modification Techniques

Since a number of activated carbon surface modification techniques are used in this study, a brief review of carbon modification techniques will be presented in this section. Although many gas or liquid phase modification methods have been known and used for many years (Puri, 1970; Kaneko et al., 1988; Ermolenko et al., 1990; Radovic and Rodriguez-Reinoso, 1997), they have not been systematically evaluated for DOM adsorption and DBP control.

Preparation of activated carbons with acidic surface

There are two main reasons for development of acidic surface functional groups on activated carbons: (i) for some specific treatment applications, such as, removal of metal cations from aqueous solutions (King, 1937; Corapcioglu and Huang, 1987; Radovic and Rodriguez-Reinoso, 1997; Dastgheib and Rockstraw, 2001) or removal of basic molecules from gas or liquid streams (Mochida and Kawano, 1991; Mangun et al., 1999b), where the goals were to enhance the surface affinity for adsorption of a particular type of adsorbate by introduction of acidic functionalities; and (ii) as an intermediate stage before introducing some other functionalities or elements to the surface (e.g., creation of carboxylic groups that act as anchoring sites for the establishment of other functional groups (Jansen and van Bekkum, 1994; Pittman et al, 1997; Cronce et al, 1997; Osmond, 2000). The latter was the main reason for using surface oxidized activated carbons in some modification experiments in this study since the presence of acidic functionalities is undesirable for the adsorption of acidic DOM molecules by GAC

Oxygen-containing acidic groups can be introduced to the carbon's surface by gas- or liquid-phase treatments (Ermolenko et al., 1990; Kutics and Suzuki, 1993; Mangun et al., 1999a). In the gas phase, oxygen, air, and oxygen-containing gases such as CO_2 and steam can be used. While low temperature oxidation favors formation of relatively strong acid groups (e.g., carboxylic), high temperature oxidation can be used to create primarily a considerable amount of weak acid groups (e.g., phenolic). The degree of oxidation depends also on the reaction time and raw material characteristics. An important issue in gas phase oxidation is that a considerable change in the carbon structure may occur through either carbon etching or pore collapsing. For example, it has been shown that oxidation of an activated carbon fiber (ACF) with air at 400°C for 30 min changed the oxygen content of ACF from 4.33% to 14.8%, while oxidation with 10% air/N_2 at 600°C for 30 min increased the oxygen content to 22.6% (Mangun et al., 1999a). The high temperature oxidation caused a severe pore widening of 30%.

Compared to gas phase treatment, oxidation in the liquid phase can incorporate higher amounts of oxygen in the form of carboxylic, phenolic, hydroxyl and quinone acidic groups onto the carbon surface at much lower temperatures (Mangun et al., 1999a). Nitric acid and a mixture of nitric and sulfuric acid have been shown to be very effective oxidizing agents (Noh and Schwarz, 1990; Gomez-Serrano et al., 1997; Miyazaki et al., 2004). Nitric acid can be used to oxidize activated carbons at temperatures ranging from room temperature to the boiling point of nitric acid (~122°C) (Ermolenko et al., 1990; Mangun et al., 1999a). It should be noted that, depending on the source and structural strength of the activated carbon (e.g., coal-based vs. wood-based), long term exposure to concentrated nitric acid, particularly at higher temperatures, might result in a serious structural alteration. Tamon and Okazaki (1996) found a decrease in the surface area and micropore volume of an activated carbon oxidized by HNO_3 at its boiling temperature. The carbon surface can also be oxidized by concentrated $(NH_4)_2S_2O_8$ solution for various periods of time (Moreno-Castilla et al., 1997). As should be expected, introduction of acidic functional groups can decrease the pH_{PZC} of carbon surfaces.

Preparation of activated carbons with basic surface

Basicity of activated carbons is associated with: (i) resonating π electrons of carbon aromatic rings that attract protons, and (ii) basic surface functionalities (e.g., amine groups) that are capable of binding protons. Basic oxygen-containing groups such as chromene- and pyrone-type functionalities can contribute to the basicity of activated carbon but their contribution is less significant as compared to the π electrons of graphene layers (Montes-Moran et al., 2004). In the presence of a large number of acidic groups (mostly oxygen-containing) on carbon surface, the contribution of resonating π electrons to carbon basicity is overshadowed. Therefore, one way of increasing basicity is to remove or neutralize the acidic functionalities, while the other is to replace acidic groups with basic nitrogen-containing groups.

Heat treatment. Removal of acidic surface groups can be accomplished by heat treatment of materials under N_2, He or H_2 atmosphere (Kaneko et al., 1988; Menendez, 1996a and 1996b; Menendez et al., 1995). The majority of oxygen functionalities on a carbon surface can be decomposed at 800 ~ 1000°C (Puri, 1970; Figueiredo et al., 1999). Hydrogen is the preferred gas since it is more effective in removing oxygen functionalities. It also stabilizes the carbon surface by deactivation of surface active sites located at the edges of the crystallites by forming stable C-H bonds. The resulting material usually shows a low degree of reactivity with oxygen or other chemical agents and a high degree of basicity (Mendendez et al., 1995; Mendendez, 1996a and 1996b). The temperature of heat treatment is usually limited to ~1000°C (Kaneko et al., 1988; Kutics and Suzuki, 1993; Mendendez et al., 1995; Mendendez, 1996a and 1996b). Although higher temperatures might be more effective in decomposing the functionalities, a severe change in the porosity and physical structure may occur. Menendez et al. (1996b) obtained a carbon with a basic surface at a lower temperature of 500°C in the presence of platinum under hydrogen flow. The role of platinum was to produce atomic hydrogen, which spills over onto the carbon surface. The atomic hydrogen hydrogasifies the most reactive, unsaturated carbon atoms at far lower temperatures than molecular hydrogen, thus leading to surface stabilization at relatively low temperatures.

Ammonia treatment. For incorporation of elements other than oxygen into the carbon structure, it is usually recommended to first oxidize the carbon to develop some oxygenated anchoring sites, and then apply the desired treatment (Pittman et al., 1997; Osmond, 2000). Economy et al. (1992) treated activated carbon fibers using NH_3 at elevated temperature to produce a fiber with basic characteristics and high adsorption capacity to acidic substances. Puri (1970) reported the formation of ammonium salts as carbon samples were treated with ammonia gas at lower temperatures in the 100-400°C range, while upon treatment at higher temperatures, amide groups were formed. Vinke et al. (1994) proposed formation of amide, imide, imine, amine, and nitrile functionalities as a result of ammonia gas treatment at 200°C. However, there was no evidence of ammonium salt formation from IR spectra. Using activated carbon fibers, Mangun et al. (2001) found that the reactivity of ammonia gas with carbon surfaces and consequent development of nitrogen-containing groups increased with the oxygen content of precursor carbon. They suggested that formation of N-containing functional groups could be controlled by temperature and time to some extent. Higher temperature (600-800°C) and longer treatment (from 10 to 120 minutes) shifted the functionalities from amine and amide to imine and imide and finally to nitrile and pyridine. Formation of these functionalities along with pyrrole and lactam, resulting from amination or amoxidation of the carbon surface, was also proposed by other researchers (Jansen and van Bekkum, 1994 and 1995). They indicated the predominant formation of nitrile-, pyrrole-, and pyridine-type functionalities upon high temperature treatment. An important feature of oxidation followed by ammonia treatment is that it is possible to enlarge carbon pores while increasing surface basicity and the number of nitrogen-containing functionalities, which is called the "etching effect" (Mangun et al., 2001). These are desirable changes in the carbon pore structure and surface chemistry for enhanced DOM removal.

Various other approaches have been reported for incorporating nitrogen functionalities on activated carbon surfaces, including chemical vapor deposition with pyridine (Kaneko et al, 2000; Yang and Kaneko, 2001; Yang et al., 2001); nitration followed by aqueous ammonia treatment in the presence of $Na_2S_2O_4$ (Abe et al., 2000), oxidation by nitric acid followed by reaction with tetraethylenepentamine (Pittman et al., 1997), and NH_3 or HCN (Stohr et al., 1991).

Impregnation of activated carbons with iron

Results obtained during the development of enhanced coagulation guidelines for drinking water treatment indicated that more DOM was removed from water using iron than aluminum coagulants (Edwards, 1997; Randtke and Hoehn, 1999). Data indicated that some DOM fractions not removed by alum could be removed by iron chloride. It was postulated that iron species may increase the electrostatic attractions to DOM, and/or favorable complexes are formed between iron and DOM species. Nowack et al. (1999) reported that ferric chloride pretreatment of water lengthened GAC bed life by lowering influent TOC and by rendering the TOC that remained more adsorbable. In addition, the use of iron oxides for DOM removal from water has received more attention in recent years (Korshin et al., 1997; Bolto et al., 2001). Therefore, one hypothesis in this study was that by impregnating activated carbon with iron, the advantages of both high surface area of activated carbon and high affinity of iron species to DOM may be combined in a single adsorbent.

Carbon materials can be impregnated with iron using methods that are used for loading various metals on catalyst supports (Radovic et al., 1997). These methods include (i) ion exchange, (ii) excess solution, (iii) incipient wetness, (iv) physical deposition of vapors, (v) chemical vapor deposition, and (vi) precipitation.

Ion exchange is one of the most widely used methods for metal impregnation. Activated carbon is equilibrated with a solution of metal ion salt, to exchange surface protons with the metal ions (Ross et al., 1985; Richard et al., 1989; Scardi and Antonucci, 1993; Pinna et al., 1995). Ferric chloride or ferric nitrate aqueous solution is usually used (Sekine and Fujimoto, 2003). This method is very sensitive to the nature of the carbon surface. The availability of a sufficient number of appropriate acidic groups (e.g. carboxylic) is a crucial factor. Therefore, development of these functionalities prior to the ion exchange stage of the process is highly recommended. After the ion exchange step, carbons are washed and filtered to remove those metal cations not strongly-adsorbed on the carbon surface. A highly dispersed metal can be loaded on the carbon using this method.

Excess solution is a common and easy impregnation method. Carbon is mixed with a solution of metal precursor, then the solvent is evaporated, or the carbon is separated from the solution by decantation or filtration before drying of the sample (Aika et al., 1972; Zaidi, 1987; Shaikhutdinov, 1995). The major difference between this method and ion exchange is that the carbon is not washed before drying. Part of the metal is adsorbed on the carbon surface, and part of it may simply be retained within the pores and the space between the carbon particles. As a result, impregnated carbon may contain both strongly adsorbed (i.e., ion exchanged) and weakly adsorbed metals (i.e., metals ions retained within the pore space).

Incipient wetness impregnation is also known as pore volume impregnation. This method requires knowledge of the carbon pore volume before impregnation. The metal precursor solution is prepared at the concentration corresponding to the desired metal loading. The solution is added to the carbon drop by drop with vigorous stirring until the last drop that the entire pore volume has been filled with metal cations (Louwers and Prins, 1992; Potoczna-Petru and Krajczyk, 1995).

Introducing metal-containing organic compounds into the carbon pores followed by thermal decomposition is another way to incorporate metals into the pores. Barnes and Dawson (1994; 1995) and Barnes et al. (1992) introduced a copper(II)-containing cellulosic compound into the pores, then decomposed the organics and obtained a copper impregnated catalyst. Other researchers (Schueller et al., 1993; Hutton et al., 1993) deposited nanometer-size particles by incorporating platinum at the molecular level using a similar method.

After impregnation in the carbon pores, the metal can be reduced or oxidized through various methods to obtain desired species. It has been reported that impregnation of iron species onto activated carbons surface resulted in reduced surface area (Chiang et al., 1999; Maroto-Valer et al., 2004). It is probably due to partial or complete pore blockage with oxygen complexes and metals at the entrance of micropores, which lead to a decrease in surface area.

Enlargement of carbon pore sizes

High temperature steam treatment has been used to create and enlarge carbon pores effectively during carbon activation for decades. It was suggested that steam treatment gasifies the least ordered and most exposed carbon atoms, thus increasing the pore size. The kinetics of the steam-carbon reaction has been studied extensively with different carbon-containing samples such as graphite and coke (Johnstone et al., 1952; Gibeson and Walker, 1966; Reichenberger, 1974; Umehara et al., 1983; Krebs and Smith, 1985). Recently, researchers from Penn State University modified activated carbon by thermal treatment in steam and increased the pore volume within the 5.5-63Å pore width range (Nowack et al., 2004; Rangel-Mendez and Cannon, 2005). Wu et al. (2005a; 2005b; 2005c) also obtained a carbon with more mesopore volume through steam treatment at ~800°C. Lin and Teng (2002) produced a series of steam-activated carbons with various degree of mesoporosity by controlling the activation time. It was found that both the surface area and pore volume of the carbons increased with the extent of activation and passed through a maximum at a burn-off of 43%, above which the porosity decreased with further activation due to the rupture of the pores. The mesoporous carbons produced were successfully used for wastewater discoloration.

Ammonia treatment at high temperature has been indicated as another effective way to enlarge the activated carbons (Mangun et al., 2001). However, an acidic surface, with enough oxygen-containing acidic functional groups is a prerequisite for effective ammonia etching (Dastgheib et al., 2004). Direct ammonia treatment does not result in obvious enlargement of carbon pores. Therefore, in order to effectively increase the carbon pore size, preliminary surface oxidation (e.g., nitric acid oxidation) is necessary. However, to create a high mesoporosity, steam treatment has some advantages over high temperature ammonia treatment due to: (1) its spread wide use in activated carbon preparation, (2) no need of preoxidation performed prior to ammonia treatment, (3) lower cost; (4) safety issues related to the use of ammonia, and (5) corrosion caused by ammonia in full scale production facility.

Preliminary Results

Prior to this project, the project team performed some preliminary investigations to demonstrate that commercially available activated carbons can be modified to enhance their DOM adsorption. The detailed results from these preliminary studies were reported elsewhere (Dastgheib et al., 2004). Here a brief summary relevant to this project will be presented.

Two commercial activated carbons, F400, a coal-based steam activated carbon, and MACRO, a wood-based chemically activated carbon, were selected for modification. They were exposed to different treatment routes, as described below, to produce sorbents with different physicochemical characteristics. The modified carbons were named with abbreviations describing the treatment procedures employed: (i) He: heat treatment at 900°C for 2hr under helium flow; (ii) H: heat treatment at 900°C for 2hr under hydrogen flow; (iii) 16NO: Oxidation by using boiling concentrated nitric acid for 1hr; (iv) 8N2H: treatment with ammonia at 800°C for 2hr; (v) 4N1H: treatment with ammonia at 400°C for 1hr; and (vi) Fe3E: iron impregnation by ion exchange method. For example, F400-He-16NO is the F400 carbon that was heat treated under helium and then oxidized with concentrated nitric acid. Virgin and modified carbons were characterized by surface area and pore size distribution analysis, pH_{PZC}, elemental analysis, water vapor adsorption, and acid/base uptake.

DOM samples were collected and concentrated from the influent of Myrtle Beach drinking water treatment plants in South Carolina using a reverse osmosis system described in Chapter 2. Therefore, it was possible to conduct constant-dose isotherm experiments with a wide range of initial concentrations to examine removal of DOM by modified activated carbons. Adsorption isotherms were performed at room temperature and an approximate pH of 7. After reaching equilibrium, solutions were filtered, then analyzed for UV_{254} absorbance and TOC concentration. Iron concentration was monitored for isotherm experiments with the iron impregnated carbons and no iron leaching was observed.

The impact of different activated carbon treatment pathways on DOM removal

In this section, a brief summary of results will be presented and discussed for F400 GAC. A detailed presentation and discussion of the results for MACRO can be found elsewhere (Dastgheib et al., 2004). However, at the end of this section, the overall conclusions for both F400 and MACRO GACs will be summarized.

The isotherm results obtained for Myrtle Beach raw water showed that different carbon surface treatment pathways could dramatically increase or decrease the DOM uptake of F400 carbon (Figure 1.2). For example, at an equilibrium concentration (C_e) of 5 mg TOC/L, an increase of up to 75% or a decrease of up to 70% in the DOM uptake by modified F400 GACs was observed depending on treatment. Therefore, it is evident that activated carbons can be tailored for enhanced removal of DOM from natural waters. In this section, a brief discussion will be presented about the impact of different surface treatment pathways on the Myrtle Beach DOM uptake by F400 GAC.

The Effect of Heat Treatment. Virgin F400, the precursor of various modified carbons, is a basic microporous carbon. Heat treatment under helium atmosphere (F400-He) did not significantly change the characteristics of the carbon. As a result, no significant change in the DOM uptake was observed. Hydrogen-treated carbon (F400-H) demonstrated a higher uptake. Better performance of this carbon can be explained by its higher basicity (i.e., higher pH_{PZC}) as compared to virgin F400 and F400-He.

The Effect of Ammonia Treatment. High temperature ammonia treatment of F400-He or F400-H did not have a significant impact on the characteristics of its precursor, or on the DOM uptake. As suggested by Mangun et al. (2001) and confirmed by our results, the reactivity of ammonia with carbon surface and consequent development of nitrogen-containing groups could be correlated with the precursor's oxygen content. These oxygen functionalities act as the anchor/target sites for ammonia molecules or NH and NH_2 radicals. Because both F400-He and F400-H had a small number of acidic oxygen groups, their physicochemical characteristics remained almost the same after ammonia treatment.

In order to increase the basicity and nitrogen content of the carbon, helium-treated F400 was oxidized by concentrated nitric acid. The low pH_{PZC} value of F400-He-16NO confirmed the existence of a large number of acidic surface groups including carboxylics, as previously demonstrated (Karanfil et al., 1996a). Due to the overall negative surface charge of this carbon and similar charge of DOM species at the isotherm conditions, its DOM uptake was reduced to approximately half of the F400 virgin.

As stated before, high temperature ammonia treatment of oxidized carbons is a promising method for the simultaneous development of nitrogen functionalities and mesoporosity. The characterization results indicated that F400-He-16NO-8N2H had both basic characteristics and significantly higher mesoporosity compared to the F400 virgin. As a result, its DOM uptake is increased by 75% (at C_e = 5 mg TOC/L) compared to the virgin carbon. This significant enhanced uptake is probably a combination of various factors: (i) availability of higher mesopore volume for DOM uptake; (ii) increased surface positive charge that favors electrostatic attraction of DOM species; and (iii) interactions of DOM components with N-containing surface functionalities.

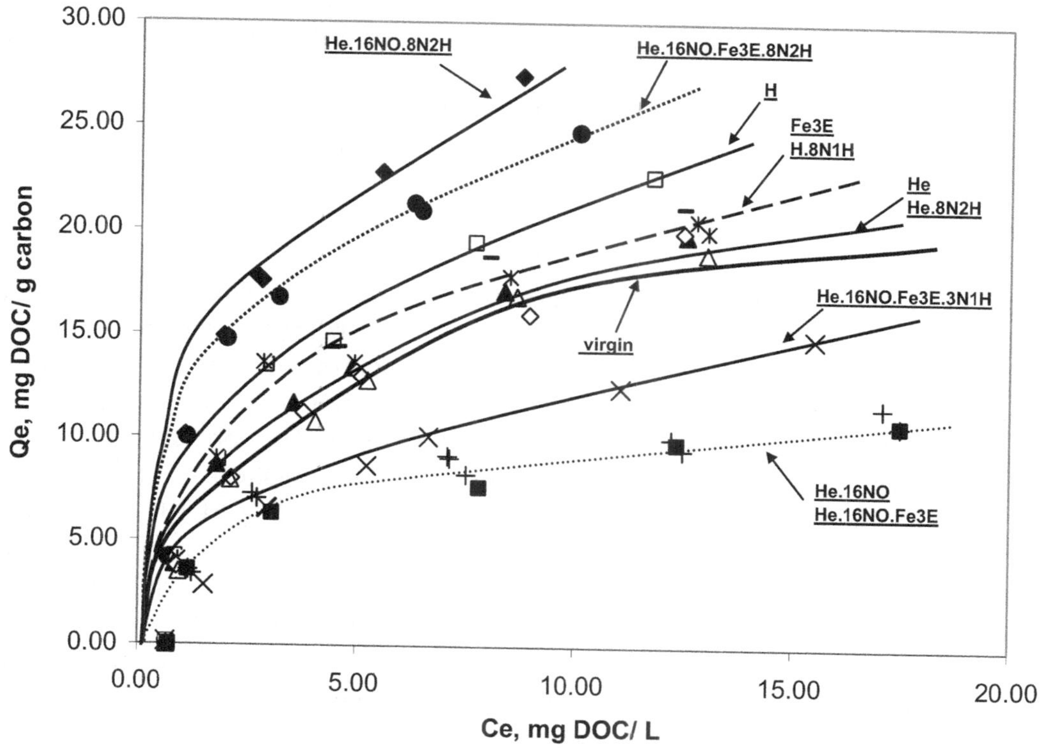

Figure 1.2 Adsorption isotherms of modified F400 GACs for Myrtle Beach raw water DOM. Lines are to show the general trends of isotherms. Since some modified F400 GACs exhibited similar uptakes, one trend line was used to depict the results for these carbons. Arrows were used to show the trend lines used for single or a group of carbons

The Effect of Iron Treatment. In order to evaluate the effect of iron impregnation on the DOM uptake, carbon precursors were treated with iron by ion exchange method. Availability of a sufficient number of suitable acidic groups (e.g. carboxylic) is a crucial factor. Therefore, development of these functionalities on the carbon surface prior to ion exchange is required. After the ion exchange, carbons were washed, filtered, and dried. A high degree of metal dispersion on the carbon surface can be obtained with this method.

F400 carbon does not have a significant number of carboxylic groups that can exchange their protons with iron cations. As a result, the ion-exchanged carbon, F400-Fe3E, had a low iron content. It is noteworthy that F400-Fe3E has acidic characteristics and one may expect a considerable drop in the DOM uptake, in comparison to virgin F400 carbon. However, its DOM uptake is slightly better than that of the F400 virgin. Therefore, the high DOM uptake of F400-Fe3E, despite its acidic characteristics, can be attributed to complexation of DOM with the iron species dispersed on the surface. Since iron content appeared to have a positive impact on the DOM uptake, it was hypothesized that DOM uptake could be further increased by using the oxidized carbon (F400-He-16NO) for ion exchange treatment and increasing the iron loading. F400-He-16NO-Fe3E, despite its high iron content, showed a DOM uptake lower than that of F400 virgin. This is probably due to highly acidic characteristics of the carbon. Therefore, it appears that it is important to optimize surface acidity and iron content of activated carbon in order to obtain maximum benefit from iron impregnation for DOM removal. In order to increase the DOM uptake of F400-He-16NO-Fe3E carbon, several processes have been employed to: (i) increase the surface basicity; (ii) increase the mesopore volume; and (iii) introduce nitrogen functionalities to the carbon surface. The high DOM uptake was observed for F400-He-16NO-Fe3E-8N2H can be related to: (1) high basicity of the surface; (2) large number of nitrogen functionalities; (3) relatively high iron loading; and (4) sufficient mesopore volume.

Conclusions from the preliminary experiments

The primary conclusion from these preliminary investigations was that it is important to have *both* a proper pore size distribution and carbon surface chemistry for the enhanced adsorption of DOM by activated carbons. High volumes of supermicropores (i.e., >10Å) or mesopores (20-500 Å) and positively charged surfaces are required to maximize the DOM adsorption by GAC during drinking water treatment operations. The presence of acidic functionalities on the carbon surface can significantly hinder the DOM uptake.

Some of the important observations with respect to F400 and MACRO from these experiments are:

- Heat treatment at 900°C increased the pH_{PZC} of F400 from 8.5 to ~10, while it did not significantly change surface area and pore size distribution of F400. Surface area normalized isotherms showed no significant changes in the DOM adsorption of F400 and Macro carbons as a result of heat treatment. The DOM uptake of these positively charged carbons (at the pH of experiments) appears to be governed mainly by their accessible surface areas. The DOM adsorption of virgin and heat-treated F400 carbons on a mass and surface area bases were higher than those of Macro GACs, although Macro carbons had about two to three times higher accessible surface area than those of F400. The main reason for this difference was that Macro carbons had higher surface acidity than F400 carbons.
- High temperature ammonia treatment of oxidized carbons, which increased the carbon basicity, nitrogen content, and accessible surface area, significantly increased the capacity of F400 and Macro (on a mass basis) for DOM. However, the uptakes on a surface area basis were almost identical, indicating that the increase in the uptake was mainly due to a higher accessible surface area.
- Iron impregnation by ion exchange followed by high temperature ammonia treatment or by excess solution method effectively enhanced the DOM uptake about 50% to 120% (on a mass basis). The surface area normalized isotherms of iron-impregnated carbons with no acidic surfaces also showed higher DOM uptakes comparing to other similar carbons, indicating the favorable chemistry effect of laden iron species on the DOM adsorption.

CHAPTER 2
ADSORPTION OF DOM BY MODIFIED GACs

INTRODUCTION AND OBJECTIVES

Powdered or granular activated carbon adsorption has been widely used in drinking water treatment plants primarily for taste, odor and synthetic organic contaminant (SOC) removal. It is also one of the recommended technologies for removal of DOM from water (Jacangelo et al., 1995; Speth, 2001). However, it has not been widely used in practice for controlling DOM due to the low equilibrium capacities and slow adsorption kinetics of activated carbons to DOM. The main reason for these drawbacks is that the majority of commercial activated carbons have been developed primarily to remove small molecular weight hydrophobic SOCs from water. Using a series of modified and well characterized activated carbons, our recent research demonstrated that high temperature helium or ammonia treatment, and iron impregnation of a coal-based microporous granular activated carbon (GAC) (F400, Calgon Corp.) and a wood-based mesoporous GAC (Macro, Westvaco Inc.) increased the DOM removal from a water source in Myrtle Beach, South Carolina (Dastgheib et al., 2004).

The main objectives in the first phase of this study were to: (i) further confirm these previous findings using a different GAC as precursor for modification and testing DOM with significantly different characteristics obtained from two water sources, (ii) examine the adsorption of DOM before and after conventional water treatment processes by modified activated carbons, and (iii) characterize DOM before and after adsorption to gain some insight to the removal of DOM by the modified activated carbons.

MATERIALS AND METHODS

Adsorbents

A GAC (i.e., HYDRODARCO 3000 (HD3000), Norit Inc.) and two activated carbon fibers (i.e., ACF10 and ACF20H, American Kynol Inc.) were used as adsorbents. Since the previous investigation showed that carbon mesoporosity played an important role in the DOM uptake, a coal-based, steam-activated, and mesoporous GAC (HD3000) was selected as the precursor for modification. The activated carbon fibers are highly microporous adsorbents with relatively uniform pore sizes and similar surface chemistries. They were used in this study to investigate the effect of carbon pore size on the adsorption of DOM from raw and treated waters.

HD3000 was modified using some of the surface modification pathways that were described in detail in our previous study (Dastgheib et al., 2004): (1) heat treatment under helium at 900°C for 2hr (HDHe); (2) heat treatment under helium at 900°C for 2 hr followed by oxidation with concentrated nitric acid and finally ammonia treatment at 800°C for 2 hr (HDN); and (3) heat treatment under helium at 900°C for 2 hr followed by oxidation with concentrated nitric acid then iron impregnation by the ion-exchange method and finally followed by ammonia treatment at 800°C for 2 hr (HDFe). At the end of preparation, modified carbons were extensively washed with distilled and deionized water. After washing, the carbons were dried and stored in a dessicator until use. GAC particles between the 30- and 40-mesh sizes were used in all experiments. In this chapter, the abbreviation codes of HD, HDHe, HDN, and HDFe were used for virgin, heat-treated, ammonia-treated, and iron-impregnated carbons, respectively.

Characterization of Adsorbents

Various characterization methods have been used to determine physicochemical properties of the virgin and modified activated carbons. Chemical characteristics were determined by: (i) pH of point of zero charge (pH_{PZC}), (ii) total HCl and NaOH uptake, (iii) elemental analysis using XPS (X-ray photoelectron spectroscopy), and (iv) ICP-AES (inductively coupled plasma atomic emission spectroscopy) of acid-digested samples for iron content. Physical characteristics were determined from nitrogen adsorption isotherms using: (i) BET equation for total surface area, (ii) Dubinin-Radushkevich (DR) equation for micropore volume, (iii) nitrogen uptake at the relative pressure of 0.98 for total pore volume, and (iv) Micromeritics' DFT (Density Functional Theory) software for pore size distribution. Brief description of these methods are provided below

pH of point of zero charge (pH_{PZC})

One-tenth molar of NaCl solutions having different pH values (2-11) were prepared using distilled and deionized water that was boiled for removal of dissolved CO_2. Solutions of 0.5N HCl or NaOH were used in pH adjustment. One hundred milligram of carbon samples were contacted with 20 mL of 0.1M NaCl solutions with different initial pH values in 25mL vials. Blanks with no carbon were also run with the samples. Duplicate experiments were performed for randomly selected samples. Sealed vials were shaken for 48 hr at 200 rpm at room temperature, and then left for 6 hr for settling of carbons. The final pH of the solution was measured. The pH_{PZC} was determined as the pH of the NaCl solution that did not change after contact with the samples. In most cases, it was required to interpolate between two measured pH values to find the exact pH_{PZC}. Replicate tests showed that the reproducibility of pH_{PZC} values was in the range of $\pm$ 0.2 pH units.

For iron-impregnated carbons, leaching of iron to the solution (at low pH values) may introduce error to the pH_{PZC} tests. Monitoring the iron concentration in solution during these experiments showed a significant leaching below pH 3. Therefore, the most acidic solution used for the pH_{PZC} determination of iron-impregnated carbons had a pH of 4.

Total HCl and NaOH uptakes

The concentrations of total acidic and basic groups on carbon surfaces were determined from NaOH or HCl uptake. Several 25 mL vials, each with 100 mg of carbon sample, were prepared and filled with 20 mL of 0.05 N NaOH or 0.05 N HCl. Vials without carbons were also prepared and served as blanks. Samples and blanks were shaken at 200 rpm for 48 hr at room temperature, and then left for 6 hr for the settling of carbons. Ten milliliter of the solution was titrated with 0.05 N of either NaOH or HCl solution. The difference between the NaOH or HCl consumed by the blank and samples was calculated and translated to the equivalent acidity or basicity per gram of carbon. The relative standard deviation of results, determined from replicate experiments of selected samples, was less than 6%.

For iron-impregnated samples, iron species (that due to the exposure to air and water are more likely to be found in the form of iron oxides or hydroxides) were partially leached to the HCl solution during the equilibration period. The leached iron ions were precipitated when the solution was titrated with NaOH. To check the effect of leached iron on the acid uptake of carbon, the same HCl uptake experiment was performed using ferric hydroxide. Although ferric hydroxide was partially dissolved in the HCl solution but due to later precipitation during the NaOH titration, the HCl uptake was similar to that of the blank solution (i.e., close to zero). Therefore, iron leaching did not have an impact on the HCl uptake values reported here, representing the amount of total basic groups on the iron-impregnated carbon samples.

Elemental analysis

Surface elemental analysis (C, N, O) of selected samples was performed using XPS technique by the Materials Characterization Laboratory of Pennsylvania State University. The analyzed spot size was 700μm × 350μm, and the average depth of analysis was estimated as 25Å. Samples were analyzed to: (1) find the elemental composition of the surface; and (2) determine ionic and non-ionic states of the dispersed iron, in the case of iron-impregnated carbons.

For the iron content analysis, around 50 mg carbon samples were digested in 10 mL concentrated nitric acid at 180°C for 30 minutes by a microwave digestion device (ETHOS Plus Microwave Labstation). The solution was, then, filtered and diluted with DDW and analyzed by Inductively Coupled Plasma Atomic Emission Spectroscopy (ICP-AES, Thermo Jarrell Ash Model 61E) for the determination of various elements including iron.

Surface area and pore size distribution

Surface area and pore size distribution of samples were determined by nitrogen adsorption isotherms from relative pressure of 10^{-6} to 1 at 77K. Surface area of samples was calculated from BET equation. The selection of appropriate relative pressure range is a crucial factor for the surface area calculation. It is desirable to cover a large relative pressure range and have a high correlation coefficient (>0.99) for linearized BET plot. However, the selected range should also satisfy two conditions: (1) it should represent a linear range of the linearized BET plot; and (2) the calculated C parameter of BET equation should be positive. In this study, the relative pressure ranges used in BET calculations were from 0.01 to 0.1 and 0.05 to 0.2 for microporous and mesoporous activated carbons, respectively. Pore size distributions of activated carbons were calculated from nitrogen isotherms using Micromeritics' Density Functional Theory (DFT Plus) software. The total pore volume was calculated from the adsorbed volume of gas near the saturation point ($P/P_0 = 0.98$). Micropore volume was calculated by using Dubinin-Radushkevich equation in the relative pressure range of 10^{-5} to 10^{-1}. By subtracting micropore volume from the total volume, total meso- and macropore volume was determined. Reproducibility of measurements was determined from triplicate data of randomly selected samples. RSD (relative standard deviation) of BET surface area, micropore volume (DR), and total pore volume were less than 3%.

DOM samples

DOM samples were collected from Charleston Hanahan and Spartanburg R.B. Simms water treatment plants in South Carolina (USA). Raw water samples (R) were collected from the influents of the plants. Treated water samples (T) were collected after the coagulation, flocculation, and sedimentation processes (C/F/S). Both (R) and (T) samples were obtained during the same sampling trip. A reverse osmosis (RO) system, described below, was used to concentrate DOM from raw and treated waters. Mass balance calculations showed that DOM recoveries of RO isolations ranged between 86 to 95% for the waters used in this study.

RO isolation of DOM

Figure 2.1 schematically depicts the field RO system used for DOM isolation in this project. The system was constructed of stainless steel components to the greatest extent possible. The characteristics of the system are summarized in Table 2.1.

During field operation, source water was obtained directly from the inlet of a drinking water treatment plant before any chemical addition. A series of pretreatment steps was initially performed before applying the water to the RO membrane. Pretreatment included: 1) a series of pre-filters to remove larger particles, 2) a (sodium-form) cation exchange softener to remove multivalent cations (mainly Ca^{2+} and Mg^{2+}) and minimize precipitation (scaling) on the membrane surface, and 3) a final cartridge filter (0.45-μm). After pretreatment, filtered and softened source water was collected in a 55-gallon high-density polyethylene tank (not shown in Figure 2.1), which was sampled for mass balance calculations. The water volume in the tank was also recorded. Then the water was emptied into a 55-gallon stainless-steel drum that serves as a feed reservoir. DOM feed solution was pumped to a high-pressure stainless-steel multi-stage centrifugal pump using a stainless steel centrifugal transfer pump. The high-pressure pump provides the necessary flow (about 4 gpm or 15 L/min) and pressure (up to 210 psig) for efficient membrane operation. Retentate was routed back to the isolate reservoir while permeate was collected in 20-liter fluorinated polyethylene jerricans, which were sampled for mass balance calculations and the volumes are also recorded. Permeate water was saved for system cleaning. The system was designed for a feed flow of about 4 gpm, a permeate flow of about 0.5 to 0.7 gpm and a retentate flow of about 3.3 to 3.5 gpm, or about 5 to 7 times the permeate flow. This ratio is recommended to provide adequate cross-flow velocity to minimize membrane fouling. Pressure gauges (P) were installed to monitor pressure in the system. A back-pressure needle valve was used to control the operating pressure in the membrane module. Most field operations employed a transmembrane pressure of about 550 kPA (80 psig), which has been found to minimize membrane fouling during our recent work (Kilduff et al., 2004).

Characterization of DOM

There are several sophisticated techniques available for characterization of DOM components isolated and fractionated from natural waters (Lenheer and Croue, 2003). However, due to limitations in sample volumes and DOM concentrations after adsorption experiments, two simple techniques, specific ultraviolet absorbance ($SUVA_{254}=UV_{254}/TOC$) and size exclusion chromatography (SEC), were used in this study. By combining both UV_{254} and TOC into a single parameter, $SUVA_{254}$ is used as an indicator of the aromatic character of the organic matter. A high $SUVA_{254}$ value (i.e., >4 L/mg-m) indicates hydrophobic nature of DOM in a sample (Edzwald and Van Benschoten, 1990; White et al., 1997; Kitis et al., 2002). The TOC concentration and UV_{254} absorbance of 0.45-μm filtered samples were measured using a high temperature total organic carbon analyzer (Shimadzu TOC-V CSH) that operates on a catalytic combustion method, and a UV spectrophotometer (Beckman DU® 640), respectively (Karanfil et al., 2003).

Table 2.1
Characteristics and specifications of the field-scale RO system

Parameter	Specification
Membrane #	
Size	AK4040TF (3.9 inch x 40 inch)
Module type	Spiral-wound cross-flow
Material	Aromatic polyamide thin film composite
Total surface area	89 ft^2
Maximum operating pressure	110 psig
Salt rejection @ 225 psig (2000 ppm NaCl)	99% (typical)
Membrane Feed Water	
Silt density index (15 min)	less than 5
pH range	3-11
Maximum temperature	10-30°C
Maximum chlorine content	less than 0.1 mg/L
Maximum flow	6 gpm
Maximum turbidity	less than 1 NTU

Osmonics, Inc.

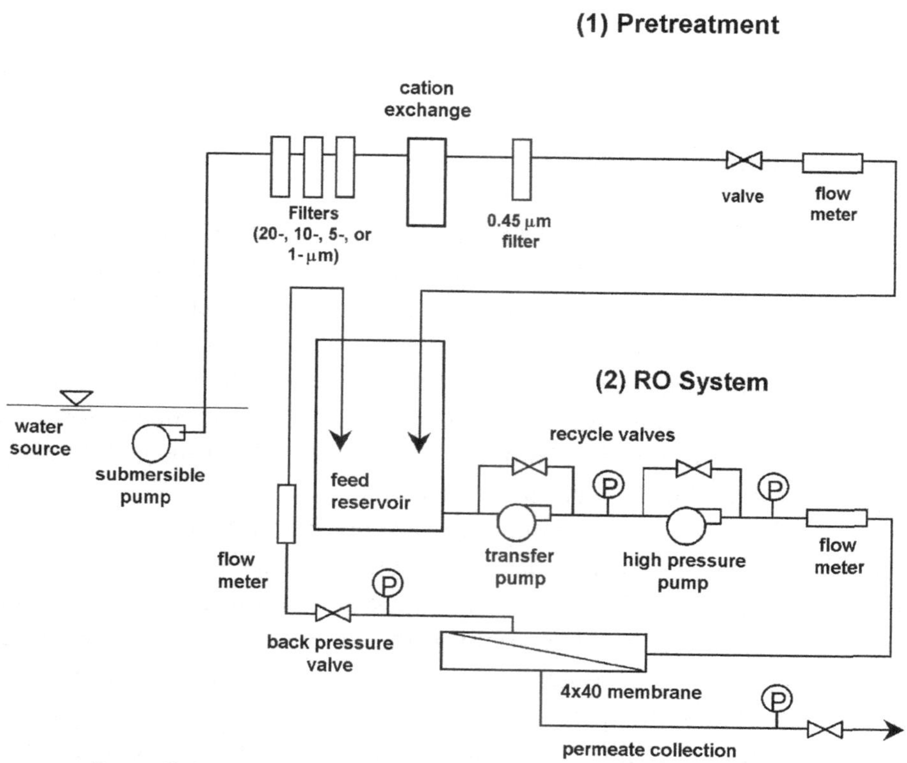

Figure 2.1 Process flow diagram of the reverse osmosis system

The apparent molecular weight distribution of DOM molecules before and after the isotherm experiments was measured using a SEC system (Dionex DX-500 with a Waters Protein Pak 125 modified silica column) equipped with a UV detector. The mobile phase was maintained at pH 6.8 by adding phosphate buffer (0.002 M NaH_2PO_4 and 0.002 M Na_2HPO_4) to 0.1 M NaCl solution. The technique employed was described in detail elsewhere (Kilduff et al., 1996; Hur, 2002). It should be noted that the SEC system used in this study provided information only about the UV-absorbing components of DOM in waters.

Adsorption Isotherms

Constant-dose bottle point isotherm experiments with 0.5 g/L carbon dose and a DOM concentration range of 2 to 60 mg TOC/L were performed for 14 days of operational equilibrium time for all water samples, following a protocol previously described (Dastgheib et al., 2004). Concentrating DOM solutions with RO allowed performing isotherms with a wide range of initial concentrations. Therefore, the impact of carbon modifications on DOM uptake could be better examined. RO concentrates were diluted to the target initial isotherm concentrations using DDW. A constant ionic strength (0.01 M as NaCl) was maintained in all DOM solutions using NaCl to eliminate the impact of ionic strength on the results. The pH of waters during the experiments remained between 6.7 and 7.6 without any buffer addition. DOM samples, collected and filtered with pre-washed 0.45-μm filters (Supor-450, Pall Gelman) before and after adsorption experiments, were analyzed for UV_{254}, TOC, and molecular weight distribution (Karanfil et al., 2003; Karanfil et al., 2005). The control bottles (i.e., carbon alone with no DOM present) packed with the isotherms indicated that there was no leaching of iron or nitrogen from modified carbons in this pH range. Some duplicate isotherms conducted during the course of the experiments showed that isotherm results were reproducible within the 10% of the reported values.

RESULTS AND DISCUSSION

Characteristics of Activated Carbons

Physicochemical characteristics of virgin HD3000, activated carbon fibers, and modified HD3000 GACs are shown in Table 2.2. The objective of heat treatment under helium was to decrease the surface acidity of the activated carbon since it has been shown that surface acidity has a detrimental impact on the DOM adsorption (Karanfil et al., 1999; Dastgheib et al., 2004). As demonstrated by the higher pH_{PZC}, higher HCl and lower NaOH uptake values, HDHe carbon has a more basic (and less acidic) surface compared to its virgin precursor. A decrease was observed in the surface area in pores >3 nm as a result of heat treatment, however this decrease constituted only 4 % of the total DFT-based surface area.

Preoxidation followed by high temperature ammonia treatment was performed to produce a carbon, HDN, with enhanced surface basicity (from decomposition of acidic groups at high temperature as well as development of basic nitrogen-containing functionalities) and increased mesoporosity (due to etching effect of ammonia). Incorporation of surface nitrogen groups was evident from the higher nitrogen content of ammonia-treated carbon compared to its virgin precursor (Table 2.2). Although the pH_{PZC} of HDN was higher than HD, it was lower than HDHe due to some residual surface acidity remaining after ammonia treatment, as indicated by the

NaOH uptake. The etching effect of ammonia treatment can be seen by the increase of the surface area in 1 to 3 nm pore size range, and micropore and mesopore volumes compared to both HD and HDHe.

Since iron forms complexes with DOM and other organic compounds and iron oxides adsorb an appreciable amount of DOM (per available surface area) (e.g., Gu et al., 1995), iron impregnation was selected as another method to increase the DOM removal of activated carbons. It was postulated that by impregnating activated carbon with iron, the advantages of both large carbon surface area and high reactivity of iron species toward DOM might be combined in a single sorbent. Characterization results showed that HDFe carbon contained about 1% iron and had a basic surface, two desired characteristics for the DOM adsorption. It is noteworthy that despite high temperature ammonia treatment, HDFe had similar nitrogen content compared to the virgin HD. However, its surface area in the pores larger > 1 nm decreased compared to both virgin HD and HDN. This indicated that nitrogen incorporation and the etching effect of ammonia were not observed in the presence of iron on the carbon surface. Therefore, it is also possible to perform the high temperature heat treatment after iron impregnation under N_2, He or H_2 atmosphere to decrease residual surface acidity on iron-impregnated carbons. XPS analyses indicated that iron dispersed on the carbon surface had oxidation states of +2 or +3. The nature of the iron on the carbon surface was not further characterized. The characterization results obtained in this study were consistent with those of modified F400 and Macro carbons produced in our previous study (Dastgheib et al., 2004).

Two virgin activated carbon fibers, ACF10 and ACF20H, have significantly different surface area and pore size distributions. While ACF10 is a microporous fiber with the majority (96%) of its surface area distributed in pores smaller than 1nm, ACF20H has an appreciable (33%) surface area in the 1 to 3 nm pore size range (Table 2.2). ACF10 and ACF20H have no surface areas distributed in pores larger than 2 and 3 nm, respectively. Both fibers have basic surfaces with only one unit difference in their pH_{PZC} values.

Characterization of DOM

SEC plots of natural raw (R) and treated (T) waters, before adsorption experiments, are shown in Figure 2.2. The apparent weight average molecular weights (MW_w) calculated from SEC data and the $SUVA_{254}$ values are provided in Figure 2.2 and Table 2.3.

The measured MW_w values are consistent with those reported for DOM in other natural waters (Chin et al, 1994; Zhou et al., 2000; Moore et al., 2001). The Charleston (R) DOM had higher MW_w and $SUVA_{254}$ values than the Spartanburg (R) DOM. This indicated that Charleston (R) water was composed of larger DOM components with higher degree of aromaticity and more hydrophobic in character compared to Spartanburg (R) water. This was in agreement with the general characteristics of these water sources. Charleston DOM originates from terrestrial vegetation that generally contains more aromatic moieties and lignin, larger MW components, and more intense color per C atom. On the other hand, Spartanburg DOM was collected from a reservoir during summer and has a more aquagenic signature, where humics with more aliphatic moieties and more carboxylic functional groups are the major contributors to DOM.

Table 2.2
Physicochemical characteristics of activated carbons and carbon Fibers

Carbon	SA_{BET}	$V_{micro, DR}$	V_{total}	$V_{meso+macro}$	pH_{PZC}	NaOH uptake	HCl uptake	DFT Surface Area Distribution (m^2/g)					Elemental Analysis*			
	m^2/g	cm^3/g	cm^3/g	cm^3/g		meq/g	meq/g	<1nm	>1nm	1-2nm	2-3nm	>3nm	O	N	C	Fe*
HD	676	0.265	0.711	0.445	6.9	0.565	0.223	353	174	81	23	70	8.7	0.7	88.4	< 0.3
HDHe	668	0.262	0.659	0.397	10.6	0.063	0.515	350	143	77	19	47	6.2	0.9	91.5	< 0.3
HDN	928	0.359	0.898	0.539	9.7	0.157	0.615	341	215	136	32	47	7.4	3.2	87.4	< 0.3
HDFe	756	0.296	0.733	0.437	8.4	0.218	0.521	391	134	88	14	32	5.6	0.7	92.7	1.0
ACF10	963	0.382	0.374	~0	8.6	0.156	0.294	1021	46	46	0	0	4.1	0.8	94.6	~0
ACF20H	1948	0.759	0.835	0.076	9.6	0.091	0.451	1055	519	494	25	0	3.3	0.6	96.1	~0

* Note: Reported elemental analysis data of oxygen, nitrogen, and carbon were obtained from XPS and represent atom% of the elements. Fe content was determined from ICP analysis of acid-digested samples and expressed as wt%.

Source: Reprinted from Cheng W., Dastgheib S.A., Karanfil T. (2005). Adsorption of dissolved natural organic matter by modified activated carbons. *Water Research* 39, 2281-2290, with permission from Elsevier.

Table 2.3
$SUVA_{254}$ and MW_w of DOM
before and after adsorption corresponding to 20 mg/L initial concentration of the isotherms

	Sample	Charleston Raw		Charleston Treated		Spartanburg Raw		Spartanburg Treated	
		$SUVA_{254}$	MW_w	$SUVA_{254}$	MW_w	$SUVA_{254}$	MW_w	$SUVA_{254}$	MW_w
		L/mg-m	Da	L/mg-m	Da	L/mg-m	Da	L/mg-m	Da
	Original Water	4.4	1716	2.0	922	2.7	1197	1.5	1074
After Adsorption	HD	3.6	1935	1.4	1039	1.6	1181	1.3	1244
After Adsorption	HDHe	3.5	1983	1.5	1130	1.7	1192	1.2	1252
After Adsorption	HDN	3.4	2016	1.1	950	1.4	1043	1.2	1221
After Adsorption	HDFe	3.3	1969	1.3	1093	1.3	1071	1.3	1230
After Adsorption	ACF10	4.5	1841	1.9	990	2.9	1126	1.5	1099
After Adsorption	ACF20H	4.9	2071	1.9	1213	2.8	1411	1.4	1336

Source: Reprinted from Cheng W., Dastgheib S.A., Karanfil T. (2005). Adsorption of dissolved natural organic matter by modified activated carbons. *Water Research* 39, 2281-2290, with permission from Elsevier.

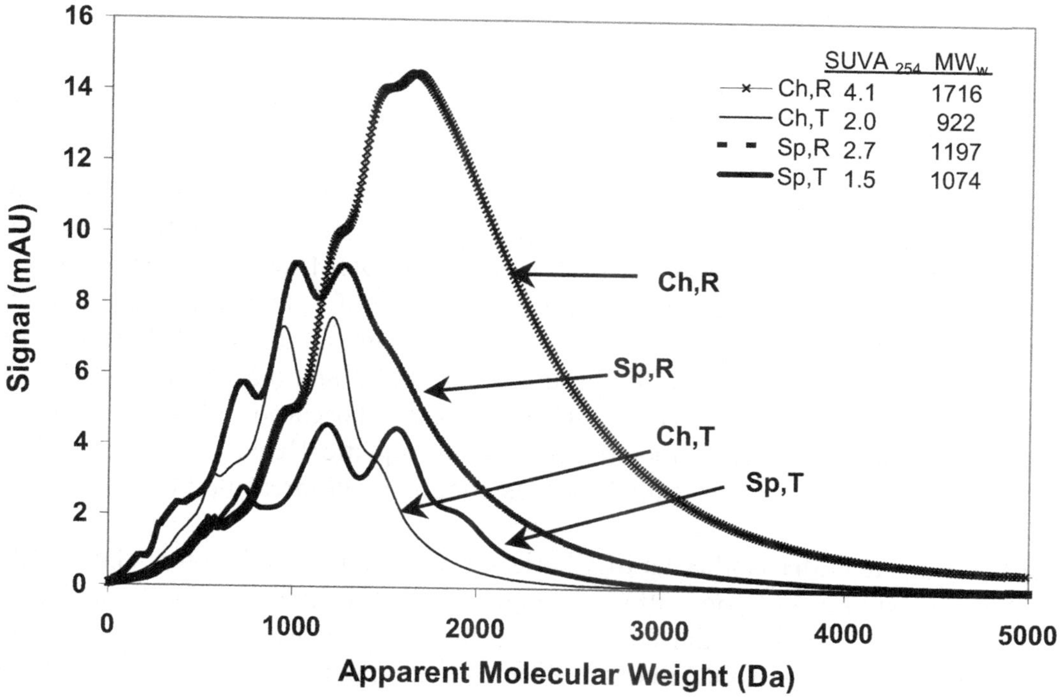

Figure 2.2 Size exclusion chromatography results of DOM used in the study. Ch: Charleston, Sp: Spartanburg, R: Raw water, and T: Treated water. The concentrations of all samples are 20 mg TOC/L. Reprinted from Cheng W., Dastgheib S.A., Karanfil T. (2005). Adsorption of dissolved natural organic matter by modified activated carbons. *Water Research* 39, 2281-2290, with permission from Elsevier.

At the time of DOM collection, the TOC removal as a result of alum coagulation in Charleston water treatment plant (WTP) was 68%, whereas it was only 37% at the Spartanburg WTP. This is consistent with the fact that the TOC removal (50 to 80%) for high $SUVA_{254}$ (e.g., 3-5 L/mg-m) waters is higher than the that (20 to 50%) for low $SUVA_{254}$ (e.g., <3 L/mg-m) waters (Letterman et al., 1999). While no prechlorination was performed prior to C/F/S at the Charleston WTP, it was being applied at the Spartanburg WTP. The characterization results indicated that C/F/S preferentially removed DOM components with higher molecular weights and aromaticity from Charleston water. This is consistent with previous observations at full-scale water treatment plants and from treatability studies (Owen et al., 1995; Vuorio et al., 1998, Matilainen et al., 2002). A higher abundance of low molecular weight components in the treated water was observed in the chromatograms presented in Figure 2.2. This is because all these chromatograms have the same TOC concentrations (i.e., 20 mg/L). Since high molecular weight fractions of DOM were preferentially removed during C/F/S, there was a higher fraction of low molecular weight components in the treated water at the same TOC concentration. In contrast to the Charleston water, MW_w decreased by only 11% as a result of C/F/S for the Spartanburg water. The chromatograms showed that the main decrease was observed in intermediate molecular weight DOM components (500-1500 Da) as a result of treatment (Figure 2.2). Despite the small change observed in MW_w, $SUVA_{254}$ decreased about 45% as a result of treatment.

Prechlorination at the inlet of the treatment plant was probably the major reason for the greater decrease in the $SUVA_{254}$ than in the MW_w of DOM in the Spartanburg water.

Adsorption Isotherms

Activated carbon fibers

DOM isotherms of activated carbon fibers are shown in Figure 2.3. For both raw and treated waters, ACF10 exhibited negligible DOM uptake. This indicated that DOM in these waters was composed mostly of macromolecules (>1 nm). The DOM uptake of ACF20H was considerably higher than ACF10, indicating that pores in the range of 1-3 nm were accessible for some DOM molecules. These findings were in agreement with (i) those of Pelekani and Snoeyink (1999) who examined adsorption of a groundwater DOM, and our previous work with the Myrtle Beach water DOM (Dastgheib et al., 2004), where similar carbon fibers were used in both studies; and (ii) the approximate diameters reported for aquatic DOM components (1-1.7 nm) in other studies having MW_w values (800-1500), as compiled and presented by Moore et al. (2001).

A comparison of ACF20H isotherms of Charleston raw and treated waters showed that within the 2 to 5 mg/L equilibrium concentration range, the DOM uptake in the treated water was approximately three to four times the DOM uptake in the raw water (Figure 2.3A). There was an appreciable amount of large DOM macromolecules (i.e., apparent MW values larger than 2000 Da) in the Charleston raw water that did not access pores less than 3 nm in width, as confirmed with SEC measurements (Figure 2.4A). The higher DOM uptake observed in the treated water was due to presence of a higher amount of low molecular weight DOM components that can more effectively access carbon pores. It was also noted that the DOM molecules with MWs larger than approximately 1600 Da in the Charleston treated water were not removed by ACF20H (Figure 2.4B). The DOM uptake of the Spartanburg raw and treated waters by ACF20H were similar (Figure 2.3B). This was attributed to the relatively small MWs of DOM before and after C/F/S. The SEC results of ACF20H before and after adsorption of Spartanburg treated water showed that DOM molecules with MWs up to approximately 2700 Da were able to access the carbon pores.

Table 2.3 shows $SUVA_{254}$ and MW_w values of DOM remaining in solution after adsorption by ACF10 and ACF20H at isotherm initial concentrations of 20 mg/L. Similar results were also obtained for initial concentrations of 10 mg/L. The $SUVA_{254}$ and MW_w values of DOM before and after adsorption by ACF10 were relatively similar, which is in agreement with the negligible DOM uptake by this carbon fiber. On the other hand, MW_w values of DOM remaining in water after adsorption by ACF20H were higher compared to before adsorption. This is a result of preferential removal of small DOM components from water, as indicated by the SEC measurements. The $SUVA_{254}$ values remained relatively unchanged after adsorption by ACF20H, suggesting that the adsorbed DOM species had relatively similar degrees of aromaticity compared to those remaining in water after adsorption.

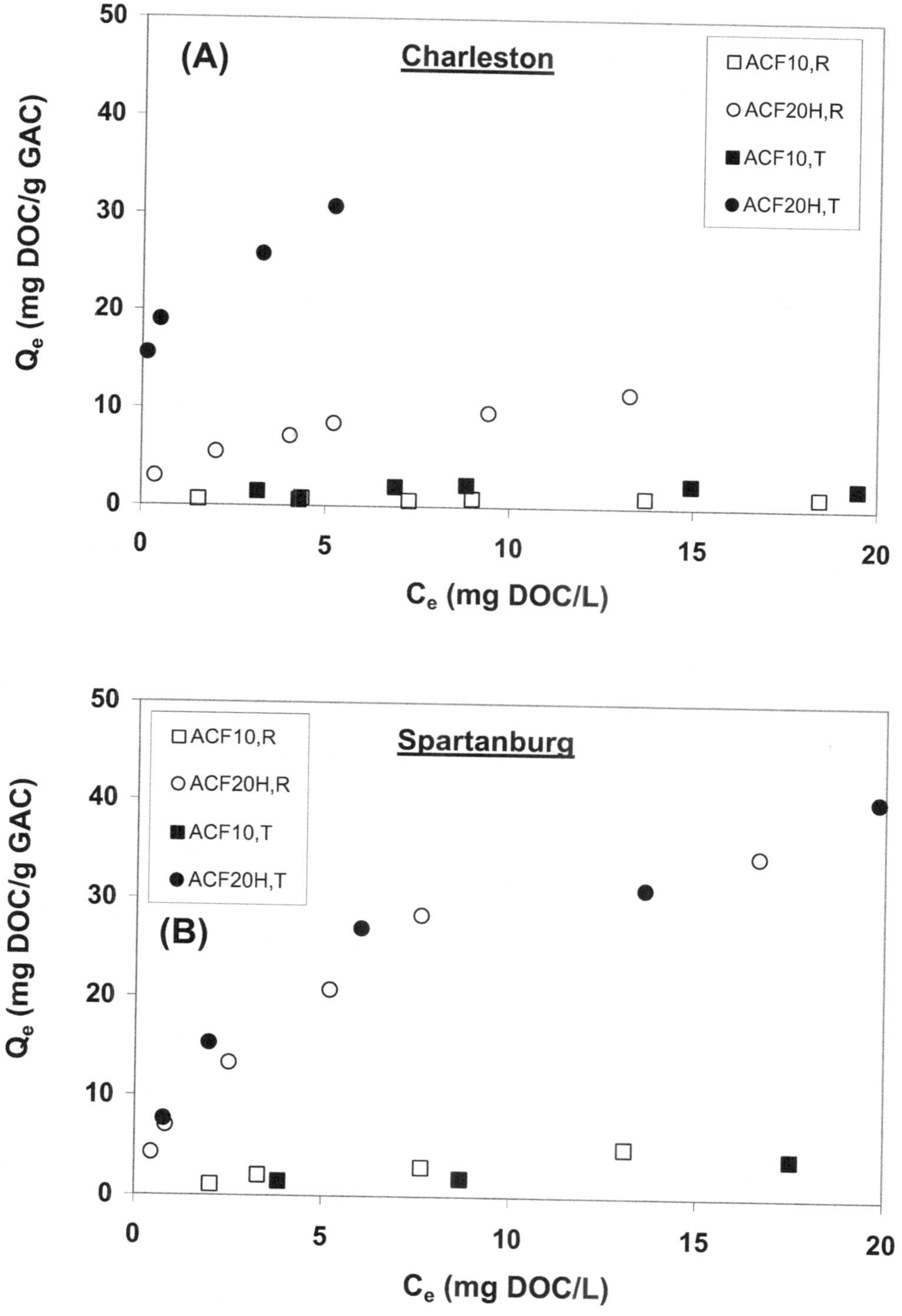

Figure 2.3 Adsorption of (A) Charleston and (B) Spartanburg DOM by ACF10 and ACF20H. R and T are to specify raw and treated waters, respectively Source: Reprinted from Cheng W., Dastgheib S.A., Karanfil T. (2005). Adsorption of dissolved natural organic matter by modified activated carbons. *Water Research* 39, 2281-2290, with permission from Elsevier.

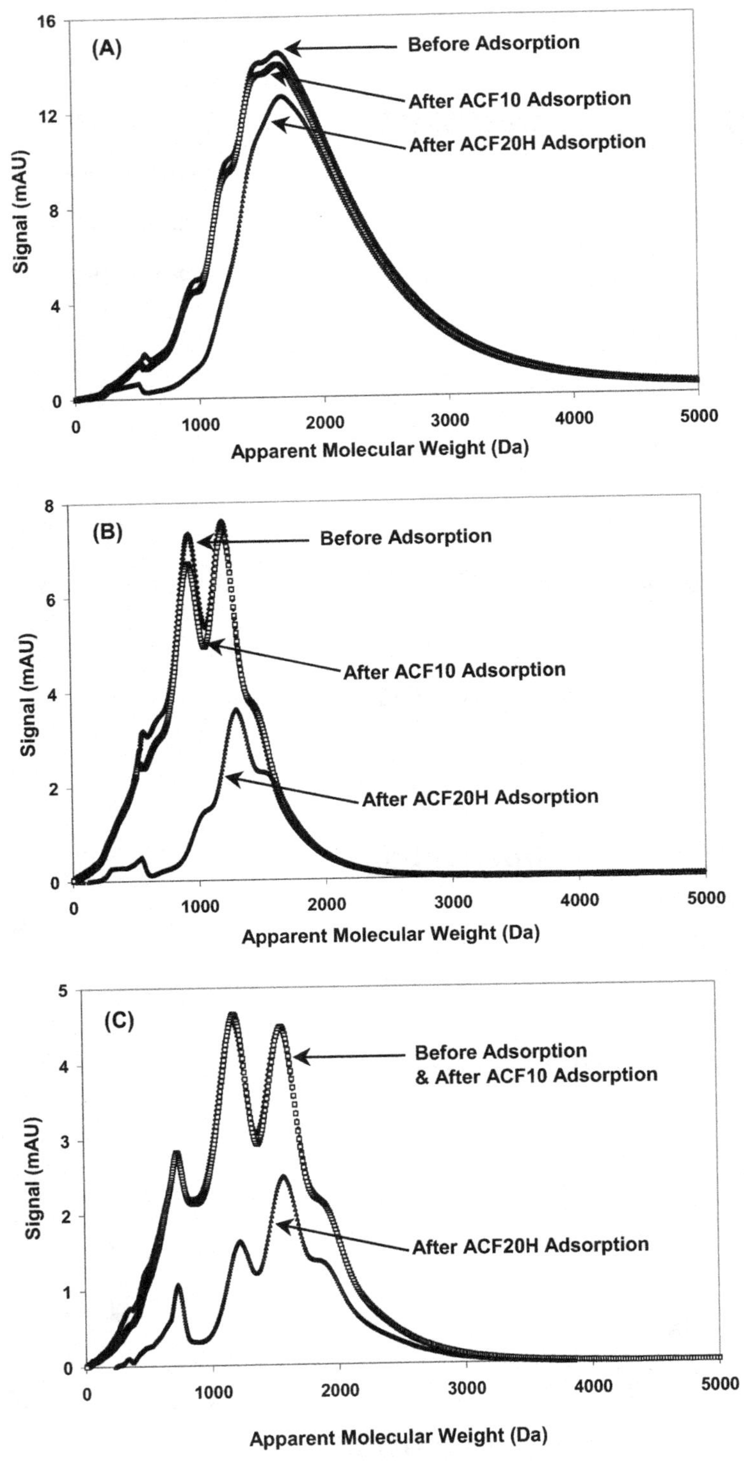

Figure 2.4 Size exclusion chromatography results for (A) Charleston raw (B) Charleston treated, and (C) Spartanburg treated waters before and after adsorption for initial isotherm concentration of 20 mg TOC/L. Source: Reprinted from Cheng W., Dastgheib S.A., Karanfil T. (2005). Adsorption of dissolved natural organic matter by modified activated carbons. *Water Research* 39, 2281-2290, with permission from Elsevier.

Modified GACs

Adsorption isotherms of Charleston (R), Charleston (T), Spartanburg (R), and Spartanburg (T) DOM by virgin and surface-modified carbons are shown in Figures 2.5 to 2.8, respectively. Each figure includes both mass- and surface area-basis isotherms for the same water. While mass-basis isotherms show the overall DOM adsorption resulting from contributions of both carbon porosity and surface chemistry, surface area-basis isotherms are used to assess the role of carbon surface chemistry. Since isotherms with carbon fibers have shown that the surface area in pores smaller than 1 nm was not accessible to DOM components, each mass-basis isotherm was divided by the accessible surface area (i.e., the area in pores larger than 1 nm) to obtain the surface area-basis isotherm.

An overall comparison of isotherms revealed similar trends for the adsorption of DOM tested in this study. On a mass-basis, the DOM uptake by the modified carbons was in the order of HDN > HDFe > HDHe > HD. On the other hand, on a surface area-basis the DOM uptake order was HDFe > HDHe ≈ HDN > HD.

Surface area-basis isotherms clearly showed that iron-treated surface had the highest DOM affinity for all waters. This was attributed to the favorable effect of surface iron species. The surface basicity of HDFe carbon was higher than the original HD. This could also contribute to enhancing DOM adsorption, which has been shown to be sensitive to the carbon surface acidity (Summers and Roberts, 1988b; Karanfil et al., 1999; Dastgheib et al., 2004). However, considering the much higher surface basicity of HDHe (pH_{PZC} = 10.6) as compared to HDFe (pH_{PZC} = 8.4), enhancement in surface affinity of HDFe was primarily from the iron effect. The total iron concentration in water before and after adsorption experiments and in isotherm controls (i.e., no DOM, carbon alone) was monitored using ICP-AES, and no leaching of iron into the solution was observed. This confirmed that the observed DOM removal from water was a result of DOM adsorption by the iron-impregnated GAC.

Surface area-basis isotherms also showed relatively similar uptakes of HDHe and HDN, which were higher than the virgin HD. The more favorable surface chemistry of HDHe and HDN can be related to their higher surface basicity (and lower acidity) (Table 2.2). More basic surfaces of activated carbon surface have more positive charges that create a higher electrostatic attraction force for the adsorption of negatively charged DOM species at the pH of the isotherms. Although a considerable number of nitrogen functionalities (as evidenced by 3.2% N content, Table 2.2) existed on HDN carbon, similar DOM uptakes of HDHe and HDN (on a surface area basis) indicated that N-groups created on the HDN did not directly improve carbon's surface affinity for the DOM removal. The total nitrogen concentration in water before and after adsorption experiments was monitored using a total nitrogen analyzer (Shimadzu TNM-1). No leaching of nitrogen was observed. This confirmed that the nitrogen on the HDN was stable.

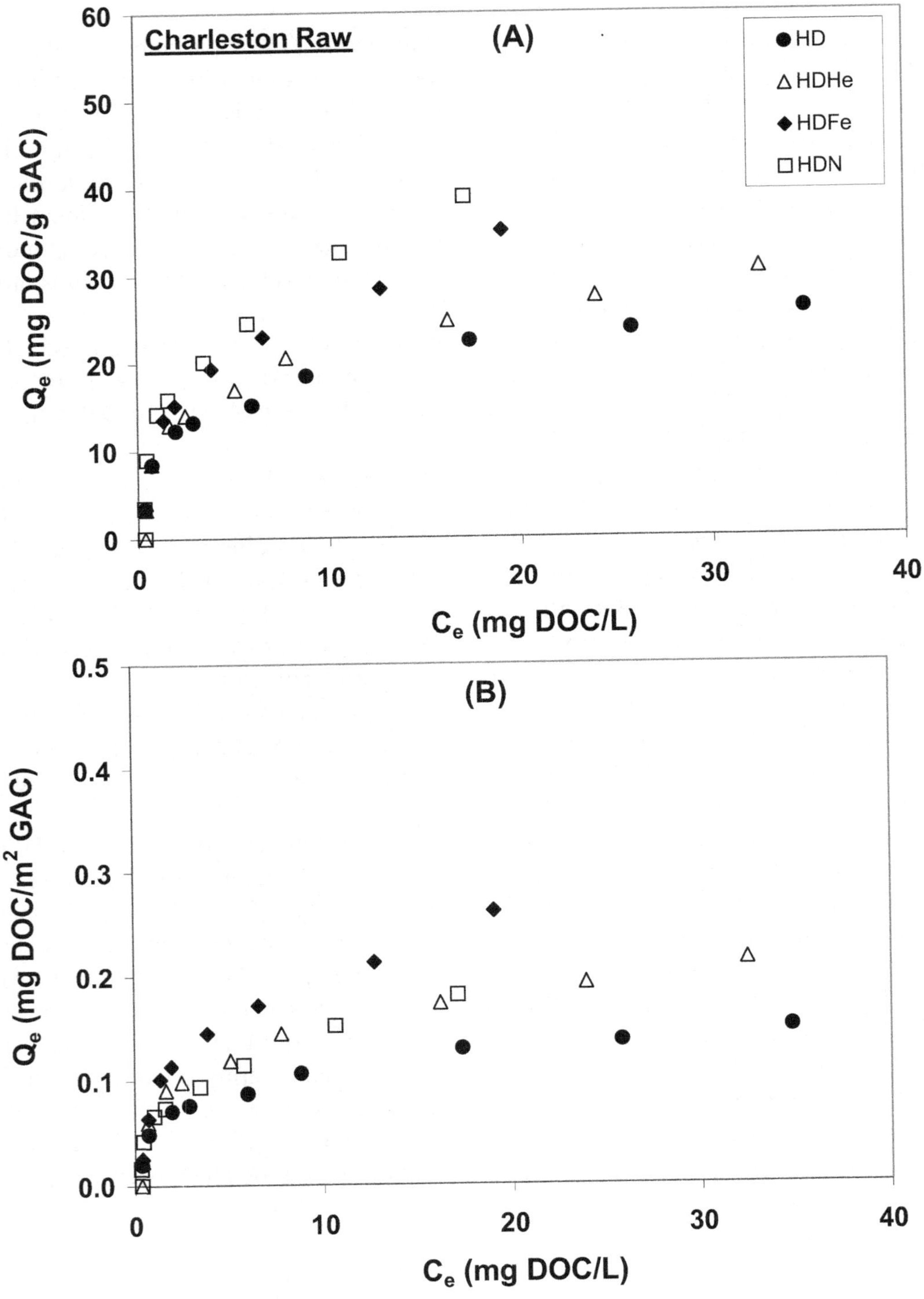

Figure 2.5 Adsorption of Charleston raw water DOM by modified activated carbons: (A) mass basis and (B) surface area basis isotherms. Source: Reprinted from Cheng W., Dastgheib S.A., Karanfil T. (2005). Adsorption of dissolved natural organic matter by modified activated carbons. *Water Research* 39, 2281-2290, with permission from Elsevier.

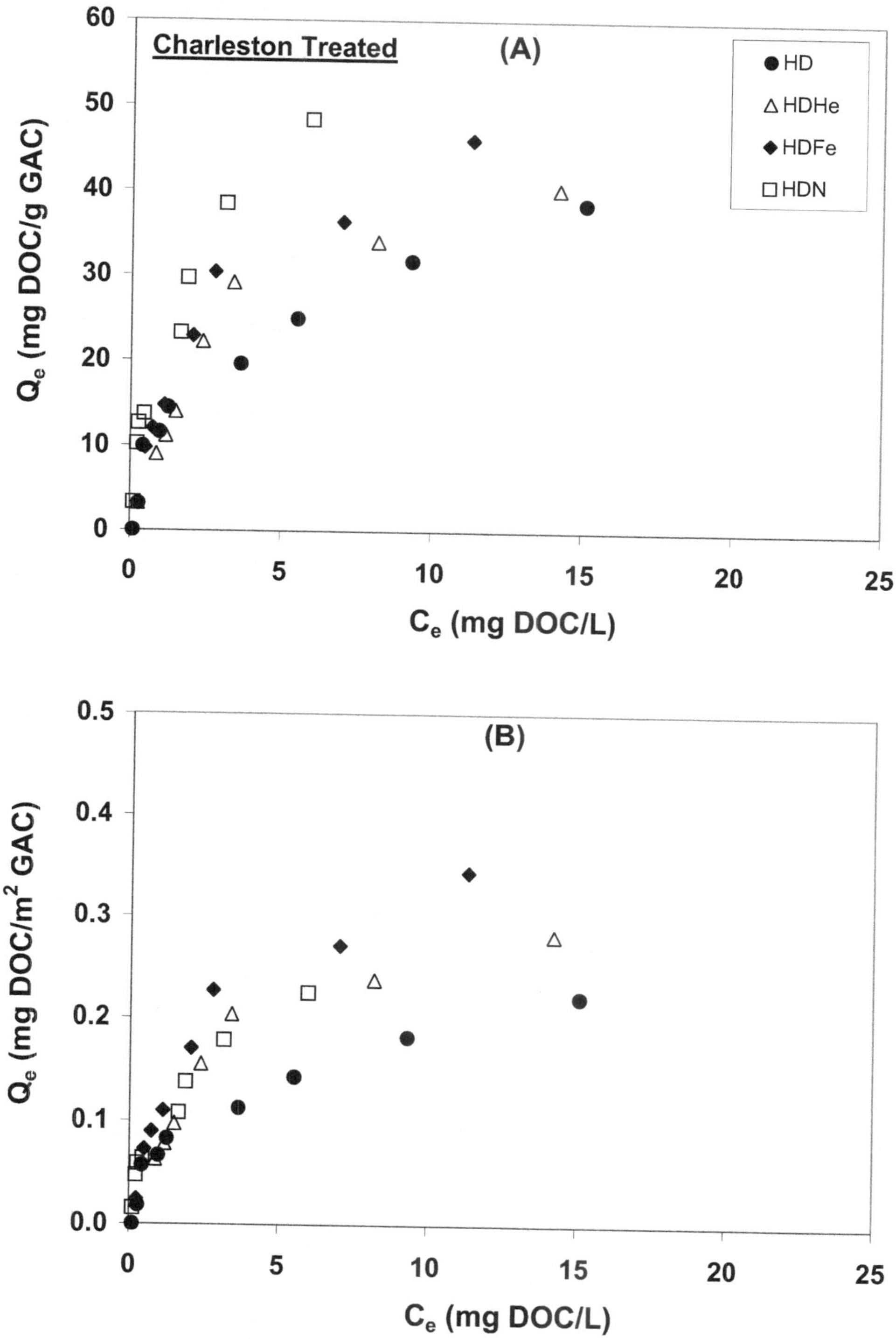

Figure 2.6 Adsorption of Charleston treated water DOM by modified activated carbons: (A) mass basis and (B) surface area basis isotherms. Source: Reprinted from Cheng W., Dastgheib S.A., Karanfil T. (2005). Adsorption of dissolved natural organic matter by modified activated carbons. *Water Research* 39, 2281-2290, with permission from Elsevier.

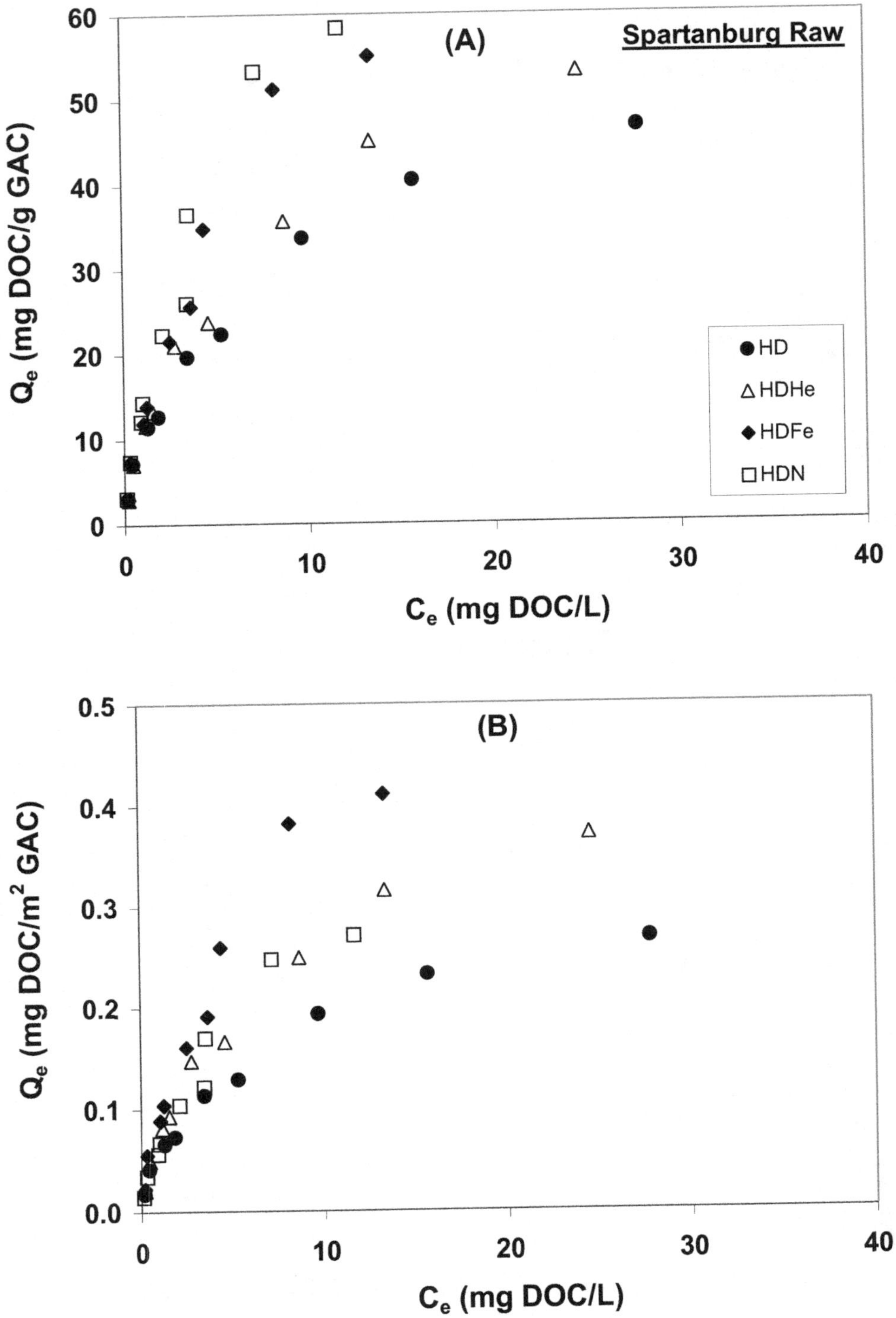

Figure 2.7 Adsorption of Spartanburg raw water DOM by modified activated carbons: (A) mass basis and (B) surface area basis isotherms. Source: Reprinted from Cheng W., Dastgheib S.A., Karanfil T. (2005). Adsorption of dissolved natural organic matter by modified activated carbons. *Water Research* 39, 2281-2290, with permission from Elsevier.

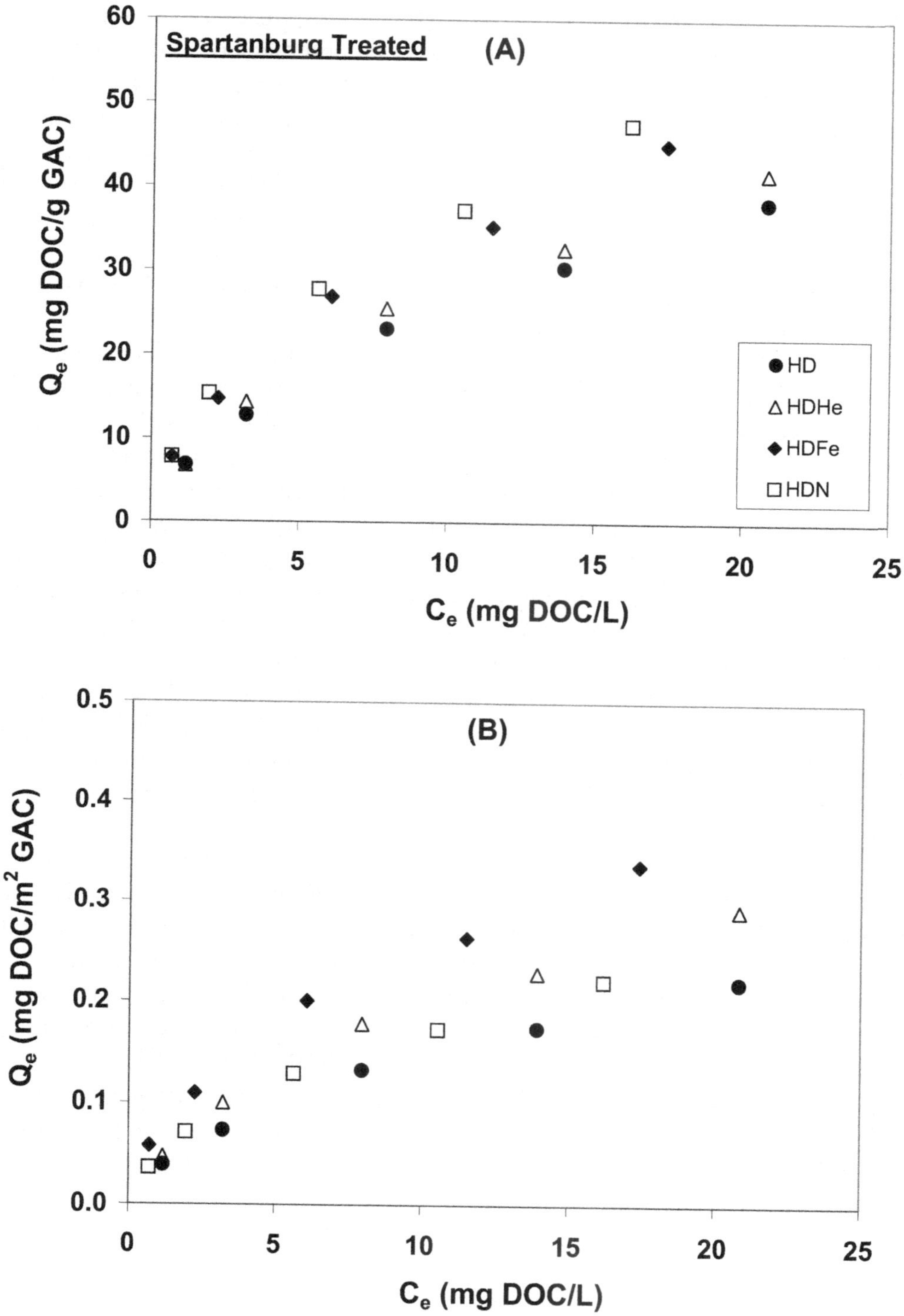

Figure 2.8 Adsorption of Spartanburg treaed water DOM by modified activated carbons: (A) mass basis and (B) surface area basis isotherms. Source: Reprinted from Cheng W., Dastgheib S.A., Karanfil T. (2005). Adsorption of dissolved natural organic matter by modified activated carbons. *Water Research* 39, 2281-2290, with permission from Elsevier.

For mass-basis isotherms, HDN carbon showed the highest DOM uptakes for all waters (Figures 2.5 to 2.8). The higher uptake of HDN over HDHe, despite its lower surface basicity, indicated the significance of accessible surface area (i.e., enlargement of carbon pores) on the DOM uptake. This observation suggests that the existence of a higher number of larger pores enhances adsorption of bulky DOM macromolecules by GAC. On the other hand, HDFe had the lowest accessible surface area among all GACs but it showed DOM uptake only slightly lower than the best performing carbon (HDN). The high DOM uptake of HDFe was attributed to the existence of iron species on the carbon surface. The mass-basis isotherms also showed slightly higher DOM uptakes for HDHe than the virgin carbon, despite their similar surface areas and pore volumes. The higher uptake of HDHe was attributed to its more basic surface, as evidenced by pH_{PZC} and acid/base uptake results (Table 2.2).

SEC and $SUVA_{254}$ measurements before and after adsorption in isotherm bottles of 10 and 20 mg TOC/L initial concentrations were used to gain some insight about the removal of DOM by modified GACs. The MW_w and $SUVA_{254}$ values of DOM remaining in water after adsorption are listed in Table 2.3 and the SEC chromatograms are shown in Figures 2.9 and 2.10. Since the trends were similar, only the results for 20 mg TOC/L were shown. The SEC chromatograms showed that DOM components from the entire MW range were removed from solution by both virgin and modified carbons. This indicated that carbon modification did not result in an apparent specific selectivity in the removal of different MW components of DOM from water. The difference between MW_w values measured after adsorption by modified and virgin GACs was always smaller than 180 Da for each DOM tested (Table 2.3). The change in $SUVA_{254}$ values as a result of adsorption indicated that more aromatic DOM components were preferentially removed by both virgin and modified carbons (Table 2.3). Similar to SEC results, the $SUVA_{254}$ values after adsorption for modified carbons were comparable to those of the virgin HD, indicating that no selective removal of specific DOM components occurred as a result of carbon modification.

It is noteworthy that although DOM uptakes of GACs were higher for treated than raw Charleston water, an opposite trend was observed for the Spartanburg water. In addition, although Charleston and Spartanburg treated DOM had similar MW_w, adsorption of the former was higher than the latter by all GACs and ACF20H. This observation can be explained by the more hydrophilic character of the Spartanburg treated DOM compared to the Charleston treated DOM, as indicated by their $SUVA_{254}$ values (Table 2.3). The higher DOM adsorption of the Charleston treated water resulted from a significant decrease in the MW_w of DOM after C/F/S. However, for the Spartanburg, raw water DOM adsorbed to a greater extent than the treated water DOM on the GACs, while they both exhibited similar adsorption on ACF20H fiber. Our characterization results are not sufficient to provide a definitive interpretation for the patterns observed for the Spartanburg water DOM.

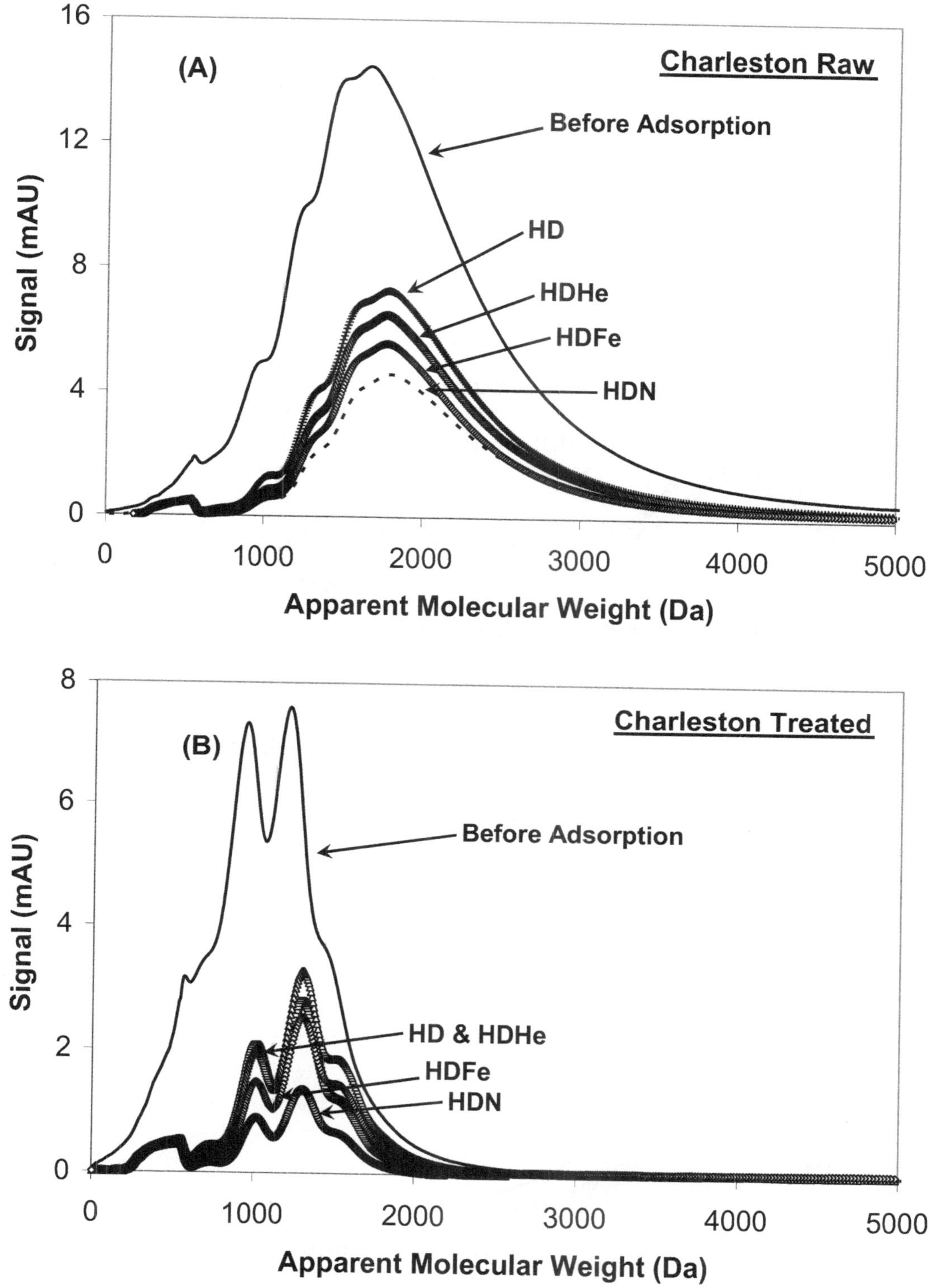

Figure 2.9 SEC of Charleston raw and treated waters before and after adsorption
Source: Reprinted from Cheng W., Dastgheib S.A., Karanfil T. (2005). Adsorption of dissolved natural organic matter by modified activated carbons. *Water Research* 39, 2281-2290, with permission from Elsevier.

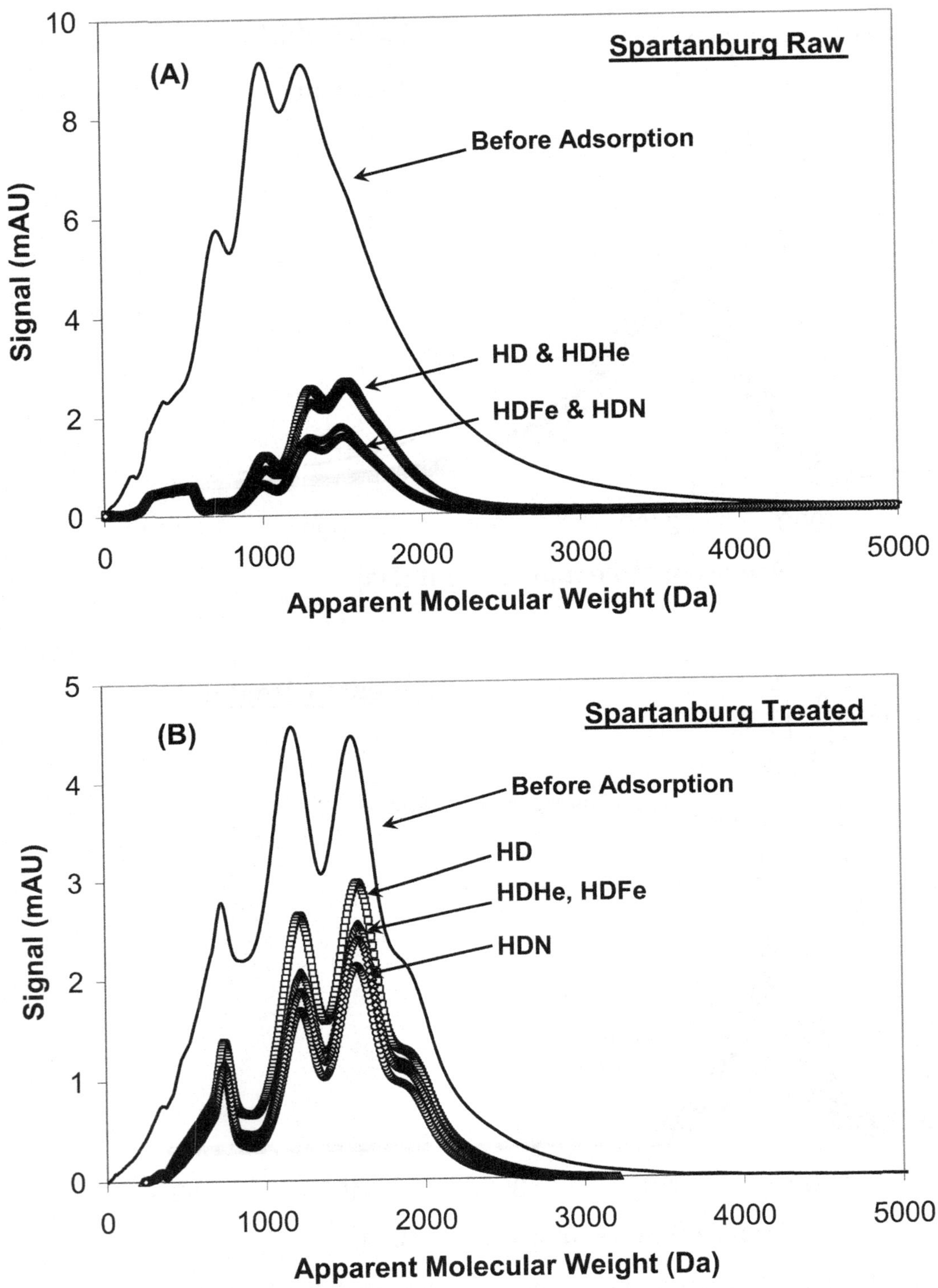

Figure 2.10 SEC of Spartanburg raw and treated waters before and after adsorption
Source: Reprinted from Cheng W., Dastgheib S.A., Karanfil T. (2005). Adsorption of dissolved natural organic matter by modified activated carbons. *Water Research* 39, 2281-2290, with permission from Elsevier.

SUMMARY OF IMPORTANT FINDINGS

The important findings from this phase are as following:

- Both carbon pore size distribution and surface chemistry play important roles in the DOM adsorption.
- Experiments with carbon fibers showed that the DOM uptake was primarily controlled by the distributions of activated carbon pore sizes and molecular weights of DOM components. The removal of DOM (with weight-average molecular weights from 922 to 1716 Da) was negligible by ACF10 having only pores smaller than 1 nm. On the other hand, small MW DOM components were preferentially removed by ACF20H, having pore widths of 1 to 3 nm. The SEC results showed that DOM components with MWs larger than 2700 Da, 1600 Da, and 2000 Da of Spartanburg treated, Charleston treated, and Charleston raw waters, respectively, were excluded from the pores of ACF20H. In contrast to carbon fibers, DOM components from entire MW range were removed from water by virgin and modified GACs.
- On a mass-basis, iron-impregnated (HDFe) and ammonia-treated (HDN) activated carbons showed significantly higher DOM uptakes than the virgin GAC. The enhanced DOM uptake by HDFe was due to the presence of iron species on the carbon surface. The higher DOM uptake by HDN was attributed to enlarged carbon pores and a basic surface created during ammonia treatment. The SEC and $SUVA_{254}$ characterization results showed no specific selectivity in the removal of different DOM components as a result of carbon modification.

CHAPTER 3
ADSORPTION OF DOM AND CONTROL OF DBP FORMATION BY STEAM-TREATED GACs

INTRODUCTION

The results obtained during the previous phase of this study showed the importance of two variables, namely the role of carbon porosity (specifically the surface area and pore volume in pores larger than 1 nm) and iron impregnation on the DOM adsorption. Therefore, the main objective in this phase was to systematically investigate the impact of activated carbon pore enlargement while maintaining basic surface characteristics on *both* DOM adsorption and DBP control.

The approach was to systematically modify a commercial activated carbon under high temperature steam treatment conditions. By increasing the treatment period, it is possible to produce progressively more mesoporous activated carbons. However, finding the optimum treatment conditions depends on the raw material characteristics. Therefore, after preliminary testing, two treatment conditions were selected to produce two steam-treated GACs having distinct and more mesoporous characteristics as compared to the precursor GAC. The effectiveness of the modified carbons for DOM removal and DBP formation control were investigated with both isotherm and rapid small scale column tests (RSSCTs). In these tests, DOM obtained from two drinking water treatment plants with low and high $SUVA_{254}$ influent waters, respectively, were used. DOM was isolated from the influent and after conventional water treatment processes (i.e., C/F/S) using the RO system. All isotherms and column experiments were performed using a single batch of water. The majority of RSSCTs were performed with the treated waters (i.e., after C/F/S) due to practical significance. The effluents from RSSCTs were chlorinated in order to assess the performance of GACs in terms of DBP formation control. In limited isotherm tests, a steam-treated GAC produced by another research laboratory was also tested.

MATERIALS AND METHODS

Adsorbents

A coal-based and steam activated carbon, HD3000 (labeled as HD, Norit Inc.), was used as precursor material for modification. Virgin GAC sample was crushed, sieved and washed with DDW, then dried at 90°C and stored in sealed containers for later experimental use. The GAC particles collected between US standard sieves of #100 and #140 mesh size (106 μm<size< 150 μm) were used in the experiments.

Approximately 150 g carbon was placed into a horizontal quartz combustion tube within a furnace. The furnace temperature was raised to 800° at 30°C/min under nitrogen gas with a flow rate of ~100 cm^3/min. Then, the gas flow was switched to high temperature steam at ~30 g H_2O/min for 2 or 4 hours. After the steam treatment, the furnace was powered off and the gas flow was switched back to nitrogen until the furnace was cooled to room temperature. The carbon sample was then washed with DDW for several times and dried at 90°C.

The two modified carbons were labeled as HD-S21 and HD-S45, respectively. S is the abbreviation for steam treatment, the number after S indicates the weight loss percentage during the treatment. Therefore, S21 and S45 indicate 21% and 45% weight loss, respectively, after the steam treatment.

In addition, a steam-treated carbon (SAC), which was produced using HD4000 (Norit Inc.), a lignite-based carbon similar to HD3000, by Dr. Fred Cannon's research group at the Pennsylvania State University was tested. This carbon was used as received with no further modification, washing or sieving. Due to limited amount available, it was only tested in the isotherm experiments. The information about the preparation of this carbon (labeled as S12L and S16L) can be found elsewhere (Nowack, 2002; Rangel-Mendez and Cannon, 2005).

Two activated carbon fibers, ACF10 and ACF20H (American Kynol, Inc.) which have relatively uniform pore size distributions and used in the previous phase of the study, were also included in the experimental matrix of this phase. No modification was applied to these activated carbon fibers. The virgin and modified GACs were characterized using the methods described in Chapter 2.

DOM Samples

For this and next phase of the project, fresh DOM samples were isolated from the influents of Myrtle Beach (MB, a high- $SUVA_{254}$ water) and Spartanburg (SP, a low-$SUVA_{254}$ water) drinking water treatment plants in South Carolina (USA) using the RO system described in Chapter 2. These new DOM isolates were used in the experiments performed in this and next chapter. The process flow diagram of Myrtle Beach and Spartanburg drinking water treatment plants are provided in Appendix A. Raw water samples (R) were collected from the plant influents, while the treated water samples (T) were obtained after coagulation, flocculation and sedimentation processes. Both plants use alum as coagulant. Mass balance calculations showed that DOM recoveries of RO isolations were 99% and 85% for Myrtle Beach raw and treated waters, and 87% and 70% for Spartanburg raw and treated waters, respectively. The important advantage of RO isolation was that all isotherms and column experiments for various carbons were performed for each water using a single batch of DOM collected from the source. Therefore, it was possible to compare the performances of various GACs side-by-side for both waters at the same TOC levels. Selected characteristics of the waters used in this study are provided in Table 3.1. The apparent weight average molecular weights (MW_w) of the DOM were determined using SEC (Figure 3.1). Since the treated waters are composed of mainly low molecular weight DOM components, higher DOM losses were observed in permeates during their isolations, as compared to the raw waters. Despite relatively similar MW_w of the Myrtle Beach and Spartanburg treated waters, less DOM was recovered in the Spartanburg (70%) than the Myrtle Beach (85%) water, suggesting that there may be some other structural differences between the DOM (e.g., molecular shape and chemistry) that were not completely resolved with SEC (with UV detection as used in this study) and $SUVA_{254}$ determination. The higher DOM recovery from the Myrtle Beach raw water as compared to that of Spartanburg is probably due to the higher MW_w of Myrtle Beach water.

RSSCTs were performed with a DOM stock solution of about 3 mg TOC/L prepared by diluting the RO concentrated raw or treated DOM stock. A typical water composition used in the RSSCTs of Myrtle Beach and Spartanburg raw and treated waters is provided in Table 3.2.

Adsorption Isotherms

Constant-dose batch isotherm experiments were performed at pH 5.5 and 7.0 using the procedure described in Chapter 2.

RSSCTs

In order to assess the performance of the produced GACs under representative drinking water treatment conditions, RSSCTs were designed following the USEPA procedure (USEPA, 1996), which is known as the proportional diffusivity approach. This procedure was shown to predict the breakthrough behavior of full-scale activated carbon columns successfully (Summers et al., 1995). The details of the RSSCT design are provided in Appendix B. The experimental set-up is shown in Figure 3.2 and its characteristics are summarized in Table 3.3. Before running each RSSCT, one hundred bed volume of DDW were passed through the column to flush the entire system. Samples were collected and measured by TOC and UV_{254} to assure that the system was free of background organic matter contamination prior to each experiment. Then, DDW was switched to DOM solution and the experiment started. The effluent samples were collected intermittently and analyzed. The system was shut down when the breakthrough reached around 70% of influent TOC value.

Table 3.1
Selected physicochemical characteristics of Myrtle Beach and Spartanburg raw and treated waters

Parameter	Unit	Myrtle Beach		Spartanburg	
		Raw	Treated	Raw	Treated
TOC	(mg/L)	12.6	3.9	2.08	1.49
UV_{254}	(cm^{-1})	0.648	0.086	0.043	0.015
$SUVA_{254}$	(L/mg-m)	5.1	2.2	2.3	1.6
MW_w	(Da)	1077	477	670	414
Bromide	(µg/L)	68	54	<MRL	27
pH	(-)	7.0	6.3	7.2	6.7
Alkalinity (as $CaCO_3$)	(mg/L)	32	8	11	6
Calcium	(mg/L)	8	11	2	4
Magnesium	(mg/L)	2.1	2.3	1.1	1.2
Iron	(mg/L)	0.69	0.01	0.03	0.04
Aluminum	(mg/L)	0.14	0.03	<MRL	0.23
Ionic Strength	(mM as NaCl)	3.94	4.83	0.43	0.34

All values are for samples filtered using a pre-washed 0.45-µm filter (either Osmonics PES or Gelman Supor) and the average of at least duplicate samples. MRL: Minimum Reporting Level of 25 µg/L. For each water, all parameters, except MW_w, were measured in a composite sample collected directly from the source during the time of RO isolation. For MW_w determination, a 20 mg/L sample diluted from the RO isolate was used.

Table 3.2
Typical water compositions of Myrtle Beach and Spartanburg raw and treated waters used for RSSCTs

Parameter	Unit	Myrtle Beach		Spartanburg	
		Raw	Treated	Raw	Treated
TOC	(mg/L)	3.2	3.0	3.3	3.0
UV_{254}	(cm^{-1})	0.160	0.061	0.115	0.051
$SUVA_{254}$	(L/mg-m)	4.9	2.0	3.4	1.7
Bromide	(µg/L)	<MRL	<MRL	<MRL	<MRL
pH	(-)	6.8	6.4	6.5	6.5
Alkalinity (as $CaCO_3$)	(mg/L)	8	8	12	10
Calcium	(mg/L)	0.08	0.11	0.05	0.37
Magnesium	(mg/L)	0.002	0.07	<MRL	0.002
Iron	(mg/L)	0.17	<MRL	0.34	0.01
Aluminum	(mg/L)	0.03	<MRL	0.04	<MRL
Ionic Strength	(as mM NaCl)	1.67	1.67	0.95	0.95
THM[a]	(µg/L)	131	84	95	42
HAA_9 [a]	(µg/L)	159	60	68	36

All values are for samples filtered using a pre-washed 0.45-µm filter (either Osmonics PES or Gelman Supor) at minimum for duplicate measurements. MRL: Minimum Reporting Level of 25 µg/L. a: THM and HAA_9 are the amount of DBP formed after chlorinating each sample for 24-hr according to Uniform Formation Condition Protocol as described in Materials and Methods.

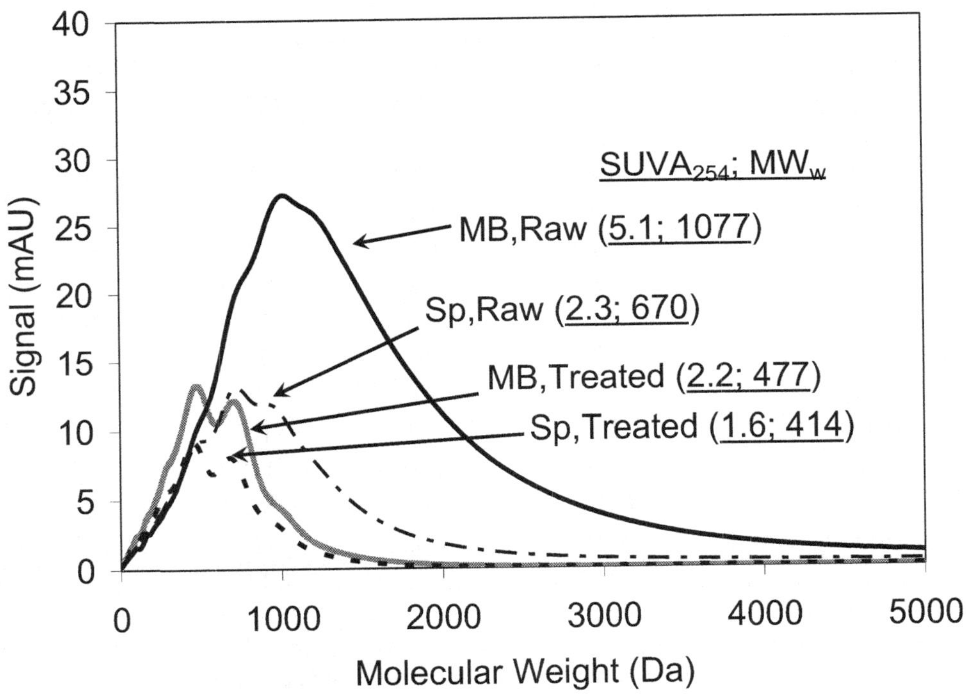

Figure 3.1 The size exclusion chromatography results of RO isolates of Myrtle Beach and Spartanburg raw and treated DOM with initial concentration of 20 mg TOC/L at pH 7

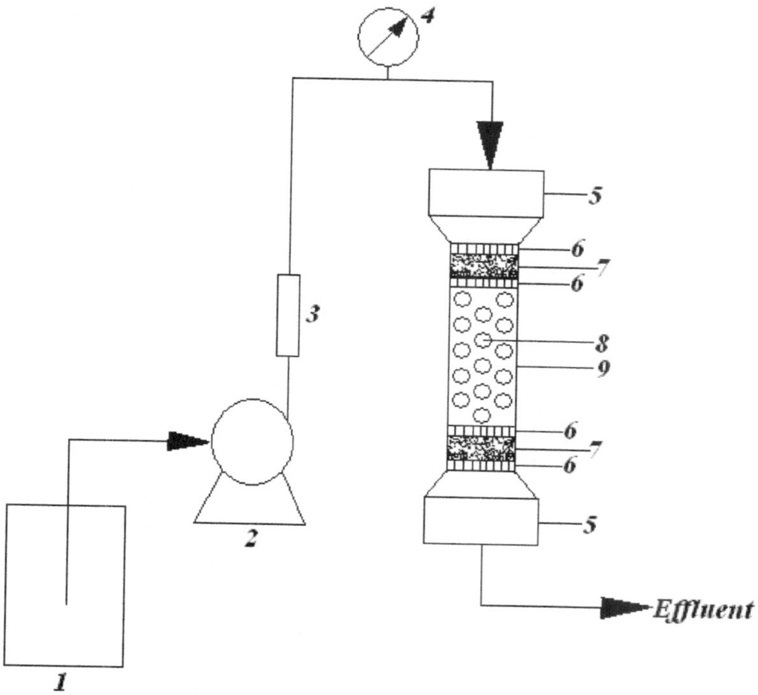

Figure 3.2 Experimental set-up for the RSSCT

Table 3.3
RSSCT design parameters

Parameter	*Value*
$EBCT_{LC}$	10 min
Scaling Factor	10
$EBCT_{SC}$	1 min
L_{SC}	15 cm
v_{SC}	15 cm/min
D_{SC}	7 mm
Q_{SC}	5.8 mL/min
$Re_{SC,min}$	1.0
d_{SC}	0.150mm (100 mesh) < d_{SC} <0.106mm (140 mesh)

It was noted that there was about 1.0 unit difference between the $SUVA_{254}$ values of the Spartanburg raw water composite sample collected for characterization during isolation (Table 3.1) and the diluted RO concentrate used for the RSSCTs (Table 3.3). No such a difference was observed for the Myrtle Beach raw water. Due to its low TOC concentration, isolation of DOM from the Spartanburg water required a significantly longer period (e.g., about a week) and processing of a greater volume of water as compared to the Myrtle Beach water, which lasted only two days. During the long isolation period, rain events occurred at Spartanburg. For the creation of a raw water composite, one sample (2L) was collected every day directly from the source at around noon. Therefore, the composite sample obtained over one week of isolation was not an ideal time or flow proportional composite sample. It is possible that the variability in the source water $SUVA_{254}$, especially during the rain events, may not have been completely captured in the raw water composites. As a result, the $SUVA_{254}$ value of RO concentrate was somewhat different from the raw water composite sample. No difference in the $SUVA_{254}$ values of the treated composite samples and RO isolate was observed.

Resin Fractionation of RSSCT Breakthrough Samples

Resin fractionation experiments were performed to determine the hydrophobic/ hydrophilic distribution of DOM components eluting from the RSSCT column. Since the sample volumes of the collected breakthrough samples were limited, a batch-mode experimental approach was developed to determine the hydrophobic/hydrophilic distribution of DOM in each sample. The approach was to contact a sample with an overdosed amount of XAD-8 resin at pH 2 for one week, allowing the removal of all adsorbable (i.e., hydrophobic) DOM fractions from water. The DOM remaining in the sample is called the hydrophilic fraction. Initial kinetics experiments showed that one week was satisfactory to reach equilibrium.

To determine the resin dose for these experiments, variable dose experiments were initially conducted using the influent solution of RSSCT. The results for the Myrtle Beach and Spartanburg water are shown in Figure 3.3. For resin doses higher than 6 g/L, the TOC concentration after adsorption remained almost constant in all samples, indicating that all adsorbable fractions were removed from solution. . Therefore, a constant XAD-8 resin dose of 8 g/L was applied to all breakthrough samples that contained DOM levels ranging from trace amounts to the influent concentration. The resin was mixed with each RSSCT breakthrough sample in 40 mL brown vials at pH 2 (adjusted by HCl) for one week. Although the resin was extensively washed by a sequence of methanol, acetonitrile and DDW prior to use, control vials that contained only the resin and distilled and deionized water were also run to monitor TOC leaching from resin. The TOC leaching during the fractionation period was less than 0.3 mg/L. The measured amount of TOC leaching was subtracted from the measured DOC values after resin adsorption. The TOC concentration of samples, filtered with pre-washed 0.45 μm filters after the fractionation experiments, was analyzed with TOC. The percentage of hydrophobic DOM in each sample was calculated as the ratio of the amount of TOC removed from water to the initial TOC concentration of the experiment.

The results of the fractionation experiments are shown in Table 3.4. As expected, high-$SUVA_{254}$ Myrtle Beach raw water had a higher hydrophobic DOM content than low-$SUVA_{254}$ Spartanburg water. Conventional treatment processes removed hydrophobic and high molecular weight components preferentially in the Myrtle Beach drinking water treatment plant where there was an average of 70% TOC removal during the RO isolation period. However, the change in DOM hydrophobicity, measured with the batch fractionation approach, as a result of conventional treatment was minimal for the Spartanburg water. This was probably due to low average TOC removal, about 29%, at this plant during the RO isolation. The SEC results indicated preferential removal of high molecular weight components in Spartanburg water. The SEC results show a wider molecular weight distribution for the Myrtle Beach DOM as compared to Spartanburg water DOM. The $SUVA_{254}$ values indicate a significantly higher aromatic character of the Myrtle Beach water than the Spartanburg water. These results are consistent with the fact that coagulation/flocculation is more effective for TOC removal in high $SUVA_{254}$ waters, and preferentially removes high molecular weight, high $SUVA_{254}$ and hydrophobic DOM components, as observed for the Myrtle Beach water. It is also evident that the extent of TOC removal, preferential removal of high molecular weight, high $SUVA_{254}$ and hydrophobic DOM components significantly decreases with decreasing the $SUVA_{254}$ of the raw water, as observed for the Spartanburg sample.

Table 3.4
Hydrophobic/Hydrophilic distribution of DOM samples used for RSSCTs

Water	$SUVA_{254}$ (L/mg-m)	Hydrophobicity (%)	Hydrophilicity (%)
Myrtle Beach Raw	4.9	54	46
Myrtle Beach Treated	2.0	37	63
Spartanburg Raw	3.4	42	58
Spartanburg Treated	1.7	39	61

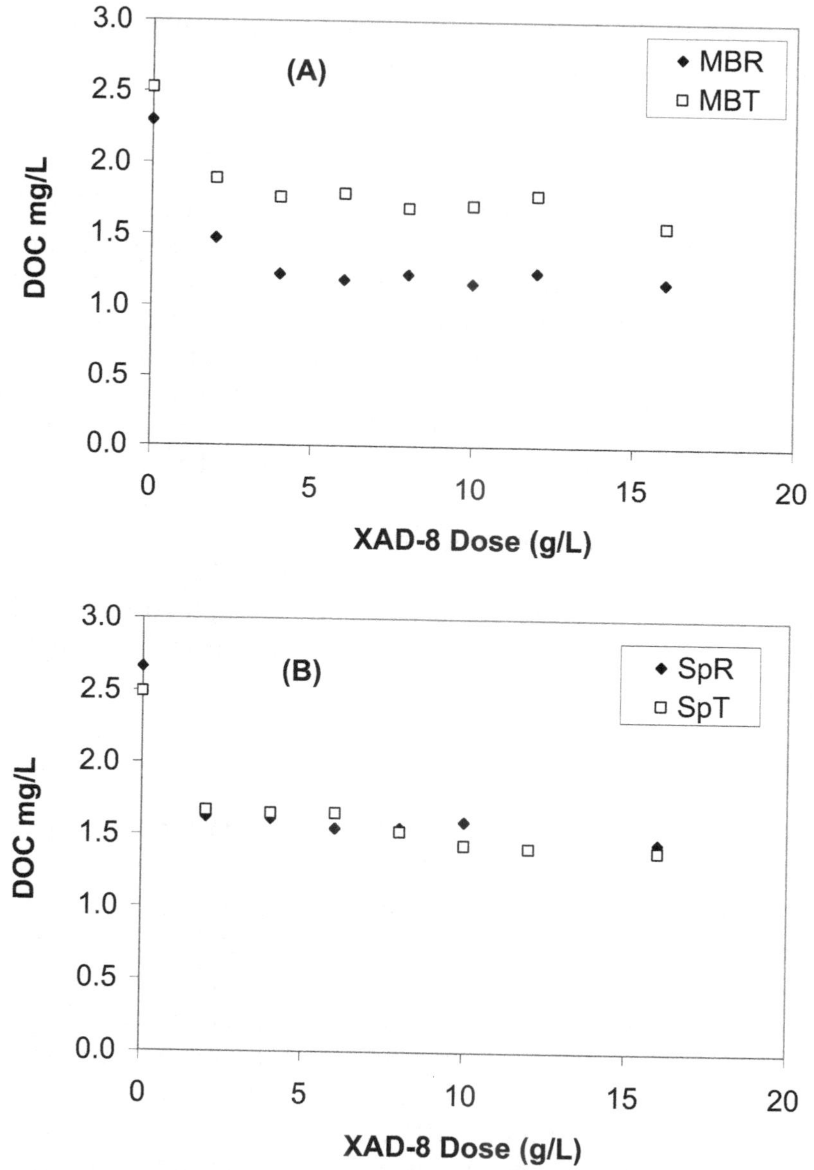

Figure 3.3 Adsorption of DOM in Myrtle Beach and Spartanburg waters as a function of resin dose

DBP Formation Experiments

THM and HAA_9 formation of RSSCT effluents were examined by chlorinating the breakthrough samples according to the uniform formation condition (UFC) protocol (Summers et al., 1996). Initial chlorination experiments were conducted to determine the chlorine dose required to obtain 1 mg/L chlorine residual after 24-hr incubation period. UFC experiments were performed using sodium bicarbonate buffer at a pH between 7.5 and 8.0. After the 24 hr incubation period, the residual chlorine was measured in the samples before quenching with stoichiometric amount of sodium sulfite (Na_2SO_3).

GAC Regeneration Experiments

Two approaches, heat treatment and desorption using NaOH, were evaluated for regeneration of steam-treated GACs. The effectiveness of GAC regeneration was examined in two cycles of batch DOM adsorption/regenration experiments.

Heat treatment

About 1 g carbon samples were placed in the heating tube equipped with Micromeritics ASAP 2010 Chemisorption Analyzer. The sample tube was evacuated and heated to 90°C at the rate of 10°C/min for 60 min to remove moisture from the carbon. Then the temperature was raised to 800°C at 20°C/min and kept for 2 hours when hydrogen or nitrogen flow was passing through the sample tube. Finally the sample tube was evacuated, and the temperature of sample tube was decreased to 600°C at 35°C/min, then to room temperature at 10°C/min. The samples was thoroughly washed with DDW and dried at 60°C overnight in the oven. The weight loss during the heat treatment regeneration was ~6% for each carbon. The carbons regenerated under H_2 or N_2 gas flow were labeled using H and N, respectively, after the carbon name. For example, HD-S21-N represents the HD-S21 carbon regenerated under high temperature N_2 atmosphere. Two different regeneration atmospheres were tested because H_2 also stabilizes the carbon surface after regeneration by deactivation of surface active sites through forming stable C-H bonds. However, high temperature H_2 treatment is practically more difficult to implement than high temperature N_2 treatment.

Desorption using NaOH

About 1g activated carbon sample that had adsorbed DOM was placed in a 250 mL Erlenmeyer flask containing 200 mL 0.5 M NaOH solution. The flask was placed on a shaker for 4 hours to desorb DOM from the carbon. Then, the carbon sample was filtered with a 0.45 μm filter paper. The filtrate was measured with UV absorbance at 254 nm. The same procedure was repeated 4~6 times until the DOM leaching from the carbon was minimal, as measured by a UV_{254} absorbance change of less than 0.001 between washes. Then the carbon was filtered and washed with DDW thoroughly while monitoring the pH of the wash water. The washed carbon was then placed in a drying oven at 60°C overnight and sealed in a container for use. The weight loss after the NaOH washing regeneration is ~10%, which includes the loss during the washing. The carbons regenerated with NaOH desorption were labeled as B after the name of the carbons. For example, HD-S21-B represents the HD-S21 carbon regenerated by NaOH desorption.

Adsorption experiments

Around 50 mg regenerated HD-S21 and HD-S45 activated carbons were contacted with 100 mL of 10 or 20 mg TOC/L Myrtle Beach treated water DOM solution in 125 mL amber bottles. The pH of the solution was kept around 7 and the ionic strength was adjusted to 0.01M NaCl. The bottles were packed and put on a tumbler for an equilibrium time of 7 days. Then the carbon samples were filtered with 0.45 μm filter paper. The carbons were collected for the next heat treatment or NaOH washing regeneration cycle. The DOM concentrations in filtered solutions were measured by TOC to determine the DOM uptake. After each carbon regeneration cycle, the adsorption experiment was repeated.

Characterization experiments

After the last regeneration cycle, activated carbons were characterized by (i) BET equation for total surface area, (ii) Dubinin-Radushkevich (DR) equation for micropore volume, (iii) nitrogen uptake at the relative pressure of 0.98 for total pore volume, and (iv) Micromeritics' density functional theory (DFT) software for pore size distribution to examine the impact of regeneration.

Analytical Methods

A number of analytical methods were used to measure various parameters in this phase and during the course of this study. These methods were developed according to USEPA or Standard Methods. The parameters, measurement methods, the instruments used and minimum reporting levels or accuracy are listed in Table 3.5. The DDW that was used to dilute or prepare samples and prepare standards had a TOC of less than 0.2 mg/L and ionic strength of less than 0.006 mM.

Table 3.5
Analytical methods and minimum reporting levels

Parameter	Unit	Measurement Method	Instrument	Minimum Reporting Level or Accuracy[a]
TOC[b]	(mg/L)	SM[c] 5310B	TOC-V CHS & TNM-1, Shimadzu Corp., Japan	0.1
UV Absorbance[d1]		SM 5910	DU 640, Beckman Inst. Inc., USA	±0.005[d2]
Total Alkalinity	(mg/L as $CaCO_3$)	SM 2320	NA[i]	2-4
Ca & Mg Hardness	(mg/L as $CaCO_3$)	SM 3500 (Ca-C and Mg-C)	ICP-AE, Spectra CIROS, USA	Ca: 0.25 Mg: 0.025
Br^-	(µg/L)	USEPA Method 300	ED 40, Dionex Corp., USA	25
Fe & Al	(mg/L)	SM 3500 (Fe-C and Al-C)	ICP-AE, Spectra CIROS, USA	Fe: 0.01, Al:0.03
Conductivity[e1]	(µS/cm)	SM 2510	M90, Corning Corp., USA	±0.5-1%[e2]
pH		SM 4500-H^+	420A, Orion Corp., USA	±0.01[f]
THMs[g]	(µg/L)	USEPA Method 551.1	6890 GC-ECD, Agilent, USA	1
HAA_9[h]	(µg/L)	SM 6251 B[h1]	6890 GC- ECD, Agilent, USA	1
Residual Free Chlorine	(mg/L)	SM 4500-Cl F	Thermo Jarrell Ash Model 61E	0.1-0.15

[a] As reported by the manufacturer.
[b] Reagent grade potassium hydrogen phthalate was used to prepare external standards. Precision ranged from 0.05 to 0.15 mg/L.
[c] SM: Standard Methods.
[d1] Measured at wavelengths of 254 nm using a 1- or 5-cm cell. [d2] Photometric accuracy (absorbance units).
[e1] NaCl standards were used to correlate conductivity readings to ionic strength. [e2] Relative accuracy.
[f] Accuracy (pH units).
[g] THM: Trihalomethanes. Methyl-tert butyl ether (MTBE) solvent extraction and a gas chromatograph (GC) with an electron capture detector (ECD) analysis.
[h] HAA_{9} 9: All nine haloacetic acid species. h1: Diazomethane derivatization, MTBE solvent extraction and GC-ECD analysis..
[i] NA: Not applicable.

RESULTS AND DISCUSSION

Characteristics of Adsorbents

Selected physicochemical characteristics of the virgin GAC, steam-treated GACs including SAC, and activated carbon fibers are provided in Table 3.6. The BET surface areas of HD-S21 and HD-S45 were 24% and 57% higher than that of HD, respectively. The total pore volumes of HD-S21 and HD-S45 were also higher than HD by 11% and 33%, respectively. The surface area in pores larger than 1 nm, which is important for DOM adsorption as to be discussed in detail in the next section, increased by 36% and 93% for HD-S21 and HD-S45, respectively, as a result of the steam treatment.

The three steam-treated carbons exhibited basic surface properties. Their pH_{PZC} values were higher than the virgin HD. The increase in pH_{PZC} can be attributed to the decomposition and removal of surface acidic functional groups, and the more aromatic carbon surface obtained after the steam treatment at 800°C. HD-S45 and SAC had, in general, relatively similar properties, except HD-S45 had a higher pH_{PZC} and surface area in pores larger than 1nm than SAC.

While ACF10 is a microporous carbon fiber with the majority (96%) of its surface area distributed in pores smaller than 1 nm, ACF20H has an appreciable (31%) surface area in the 1 to 2 nm pore size range. ACF10 and ACF20H have no surface area in pores larger than 2 and 3 nm, respectively. Both fibers have basic surfaces with only one unit difference in their pH_{PZC} values, which are comparable to the pH_{PZC} values of the steam-treated GACs.

Adsorption Isotherms

Adsorption isotherms of raw and treated water DOM from Myrtle Beach and Spartanburg drinking water treatment plants were obtained at two pH conditions (5.5 and 7.0). The results were evaluated using both mass and surface-area basis isotherms. Surface-area isotherms were used to determine the impacts of changes in carbon surface chemistry on DOM adsorption. To obtain surface-area normalized isotherms, it was necessary to determine the accessible pore size region for DOM molecules.

Determination of accessible pore size region for DOM molecules

As in Chapter 2, the isotherms of Myrtle Beach and Spartanburg DOM obtained for the two carbon fibers at pH 5.5 and 7.0 were used to determine the accessible surface area for the DOM components in the water studied. The mass-basis isotherm results are shown in Figure 3.4. The DOM uptake by ACF10 in both raw and treated waters was negligible (i.e., less than 5 mg TOC/ g carbon fiber) at both pH values. In contrast, ACF20H showed obvious uptake of DOM from both waters. Because 96% of the surface area of ACF10 is in pores less than 1 nm, this finding indicated that the pores less than 1 nm were mostly inaccessible to DOM components in both raw and treated waters.

Table 3.6
The physicochemical characteristics of adsorbents.

Sample Name	pHpzc	S_{BET} (m^2/g)	V_t (cm^3/g)	V_{micro} (cm^3/g)	$V_{meso+macro}$ (cm^3/g)	DFT Surface Area Distribution (m^2/g) <100nm	<1 nm	>1 nm	1-2 nm	2-3 nm	>3 nm
HD	5.3	686	0.712	0.266	0.446	480	335	145	59	20	66
HD-S21	8.5	853	0.790	0.325	0.465	479	282	197	81	28	88
HD-S45	10.9	1078	0.947	0.399	0.548	703	423	280	151	48	81
SAC	9.2	1027	0.905	0.387	0.518	659	453	206	133	23	50
ACF10	8.6	963	0.374	0.382	~0	1067	1021	46	46	0	0
ACF20H	9.6	1948	0.835	0.759	0.076	1574	1055	519	494	25	0

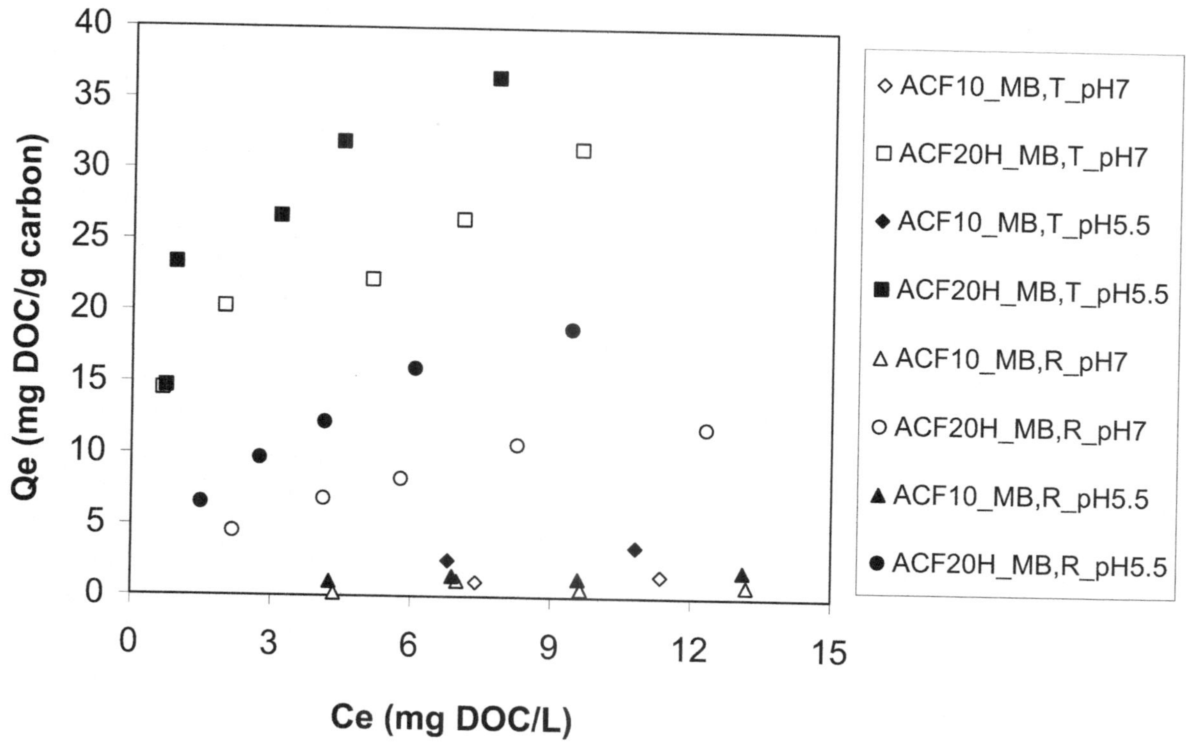

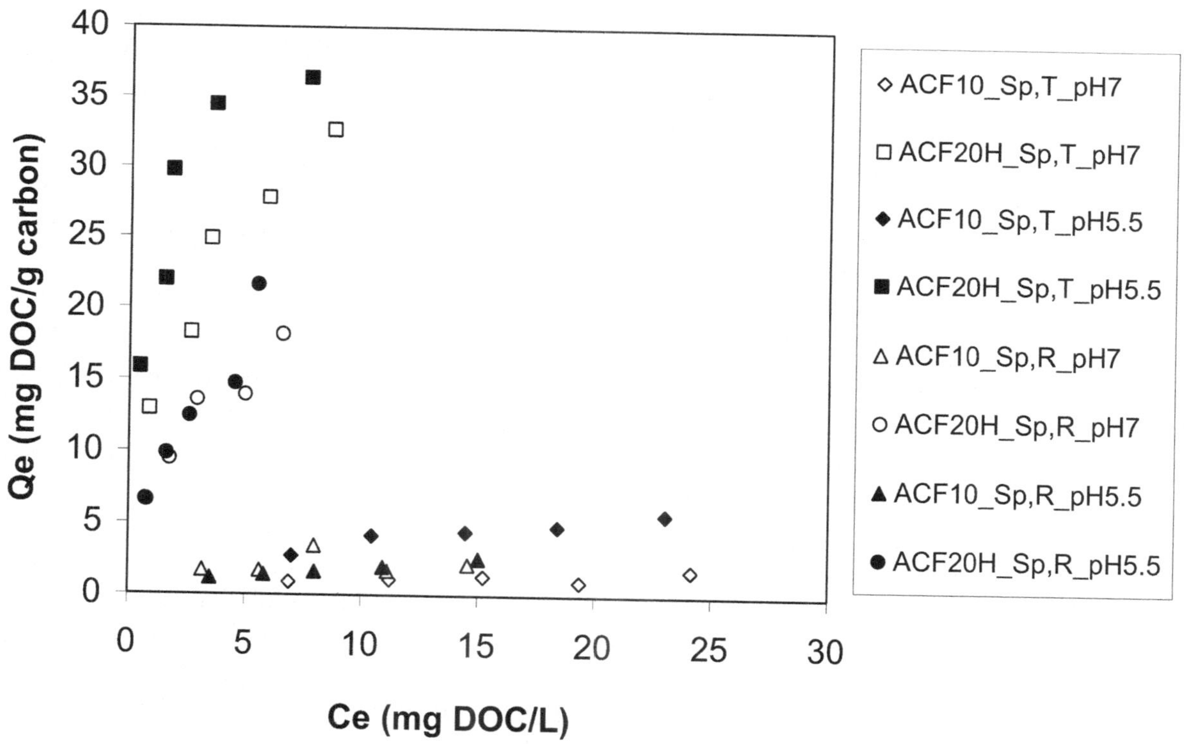

Figure 3.4 Myrtle Beach raw (MB,R), treated (MB,T) and Spartanburg raw (Sp,R), treated (Sp, T) DOM adsorption isotherms of ACF10 and ACF20H at pH 5.5 and 7

A decrease in pH increased adsorption of DOM by ACF20H in raw and treated waters. The increase in DOM uptake by GAC with decreasing pH has been previously reported (e.g., McCreary and Snoeyink, 1980; Weber et al., 1983; Karanfil et al., 1998). This has been attributed to combination of various factors with decreasing pH: (i) the more hydrophobic nature of DOM, (ii) the coiling of DOM molecules due to decreasing intermolecular electrostatic repulsion, and (iii) the more positively charged nature of the carbon surface. However, no clear pH effect was observed for the DOM uptake of ACF10. Furthermore, the treated water DOM, which is composed of mainly low molecular weight components, was not adsorbed in significant amounts by ACF10 and the DOM adsorption always remained less than 5 mg TOC/g carbon fiber at any pH for both Myrtle Beach and Spartanburg waters (Figure 3.4). This further confirmed that pores less than 1 nm were not accessible to DOM components in these waters.

Various studies have examined correlations between DOM uptake and different pore size regions of activated carbon. It has usually been concluded that secondary micropores (pores > 1nm) and/or mesopores are important for DOM adsorption (Karanfil et al., 1999; Bjelopavlic et al., 1999; Newcombe, 1999; Snoeyink and Summers, 1999; Li et al., 2003). The results obtained in this study using both raw and treated waters at the typical pH of water treatment provided direct experimental evidence that the great majority of DOM components in natural waters can not access pores smaller than 1 nm. Therefore, when selecting a GAC for DOM removal, the GACs with large surface areas and volumes in pores >1 nm should be preferred. The total BET surface area and the total pore volume are not the important parameters in selecting GACs for DOM removal because they do not provide information about the accessible pore region.

It should be also noted that the interplay between the pore size distribution in pores > 1nm and the molecular weight distribution of DOM affects the performance of GACs for DOM removal. Although the results for the steam-treated GACs will be discussed in detail in the next section, some of those results are compared here with those of ACFs to further address the accessible pore size region issue. ACF20H has a pH_{PZC} comparable to HD-S21 and HD-S45, indicating that all these carbons have relatively similar surface acidity, a factor important for DOM adsorption. ACF20H has significantly higher surface area in pores >1 nm than the other two GACs, however, the majority of this surface areas is in pores 1-2 nm. When the DOM uptakes of these three carbons are compared, ACF20H exhibited lower uptake than the other two GACs (Figure 3.5 and Figure 3.6). It is evident that the pore region of 1-2 nm, where ACF20H has a significant surface area, is not the only region where DOM molecules adsorb. For example, the size exclusion chromatography results obtained after adsorption in the previous phases of this study (Chapter 2) have demonstrated that some high molecular weight DOM components are excluded from ACF20H.

Due to heterogeneous physicochemical nature of both DOM components and activated carbon pores, it is not possible to precisely determine the *exact accessible* pore region(s) for different DOM molecules. It is, therefore, important to keep in mind that some of the data analysis that will be performed in the following sections using the surface-area basis DOM adsorption isotherms (obtained by dividing mass basis isotherms with surface area in pores > 1nm) will still have some limitations because not every DOM component can access to all pores > 1nm. Therefore, these analyses will be interpreted with some caution.

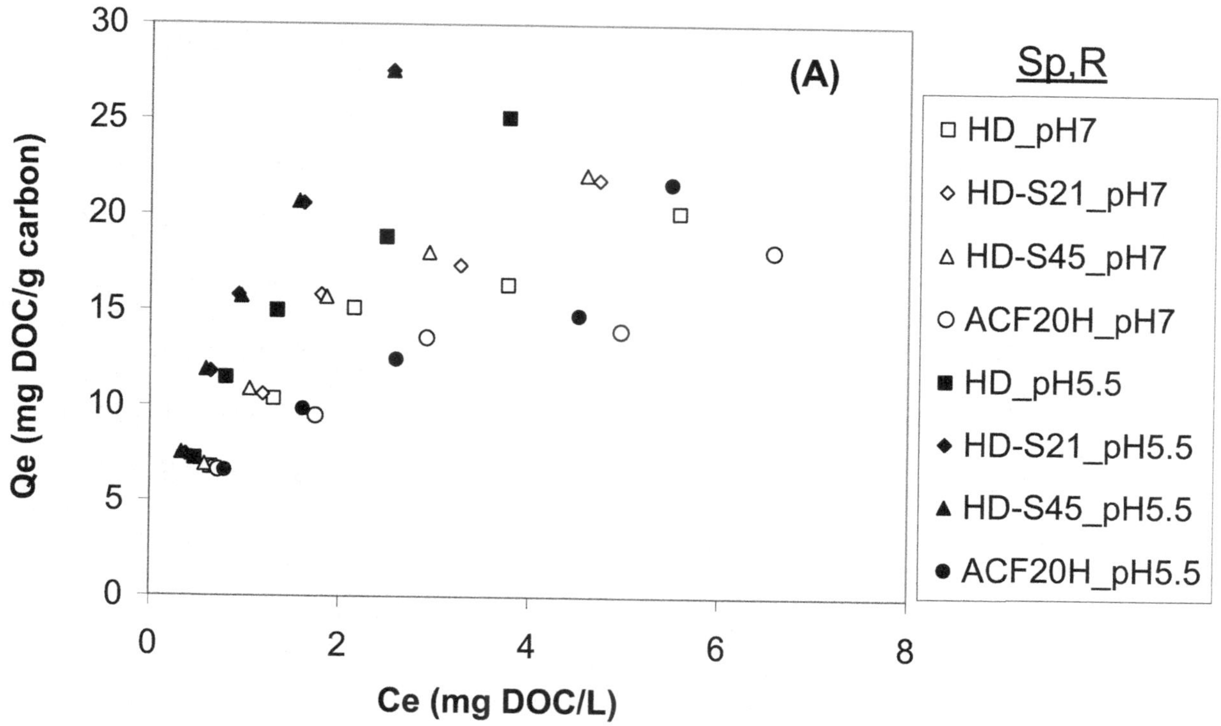

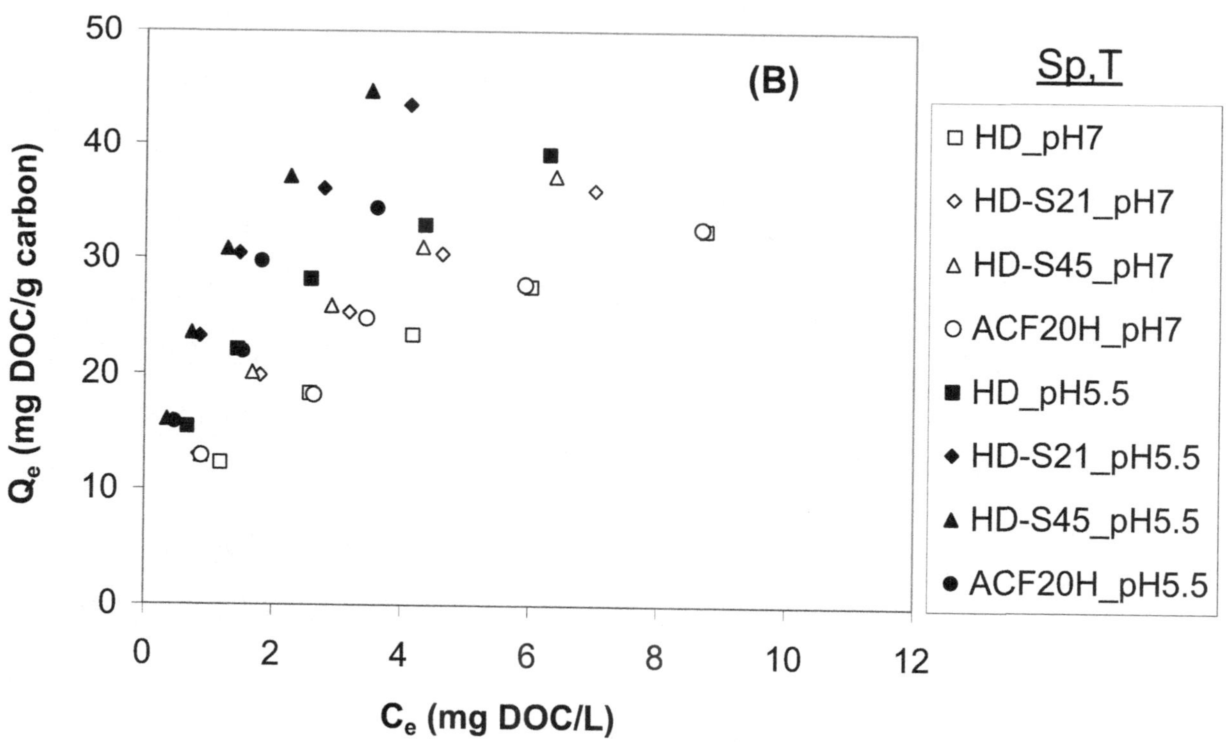

Figure 3.5 Spartanburg raw (A) and treated (B) water DOM adsorption isotherms at pH of 5.5 and 7

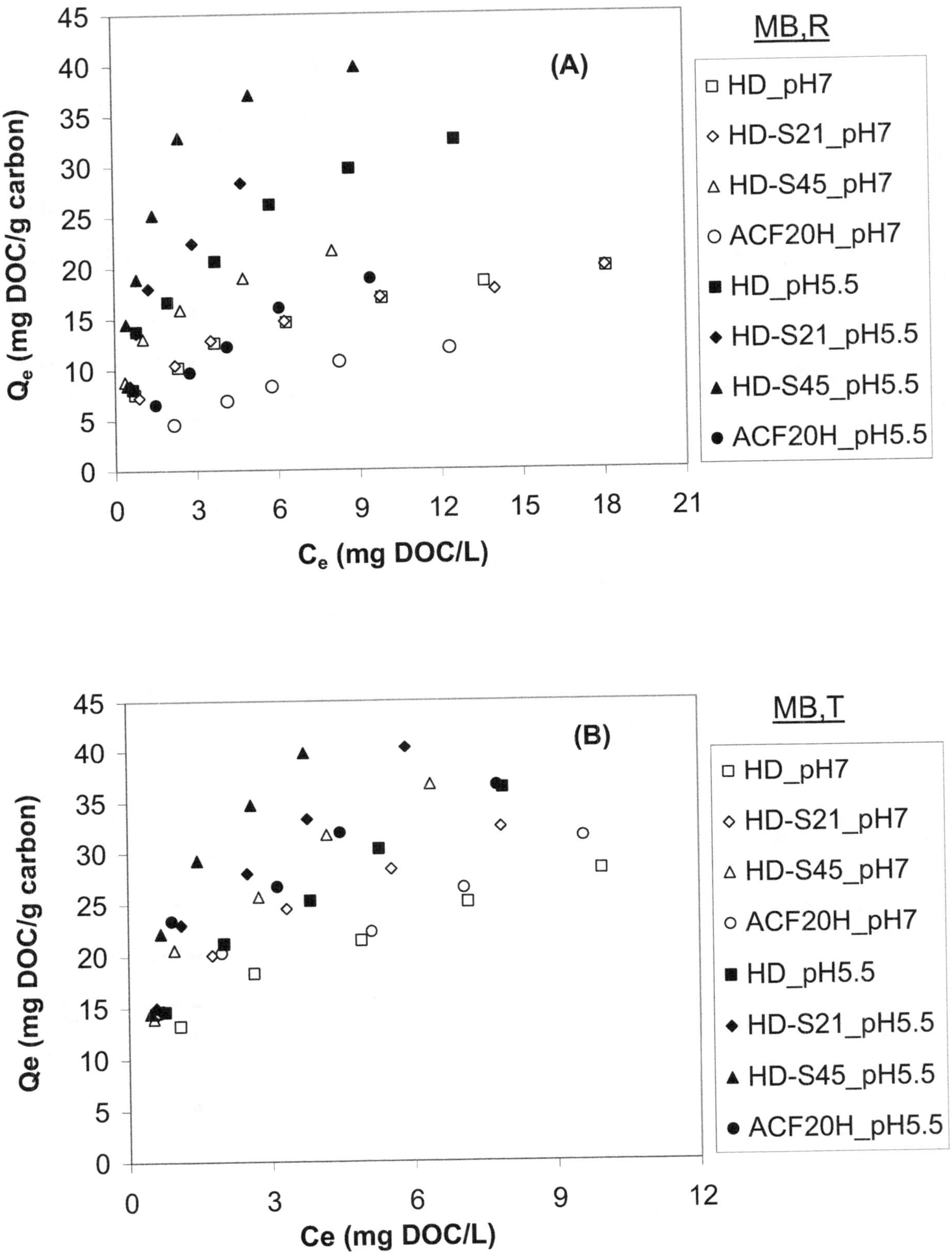

Figure 3.6 Myrtle Beach raw (A) and treated (B) water DOM adsorption isotherms at pH of 5.5 and 7

Myrtle Beach treated water

Since the treated water results are practically more relevant than those of the raw water DOM, they are presented and discussed first. Mass and surface-area basis DOM isotherms of the Myrtle Beach treated water at pH 5.5 and pH 7 by the virgin and steam-treated carbons are shown in Figure 3.7. On a mass basis, the DOM uptake of the carbons at pH 7 followed, in general, the order of SAC ≥ HD-S45 > HD-S21 > HD, while the order of DOM uptake at pH 5.5 was SAC ≈ HD-S45 > HD-S21 > HD. When carbon characterization results are examined, the observed trends suggest that the enhanced DOM uptake exhibited by the steam-treated carbons as compared to the virgin carbon can be attributed to the increase in both accessible surface area (i.e., pores larger than 1 nm) and pH_{PZC} of the carbons as a result of steam treatment. To isolate the effect of surface chemistry, the trends for the surface-area normalized isotherms were examined. It is probably more meaningful to examine the surface-area normalized isotherms of the treated water than those of the raw water because the treated water DOM has a narrower molecular weight distribution than the raw water, and it consists of mainly low molecular weight DOM components. Therefore, it is likely that a larger fraction of DOM in the treated water will be able to access larger fraction of the surface area in pores > 1 nm that was used for normalization of the isotherms. In general, for both pH 5.5 and pH 7, the surface-area basis DOM uptake followed the order of SAC > HD ≈ HD-S21 > HD-S45. This order did not correlate with the pH_{PZC} values of the carbons; HD-S45 had the highest pH_{PZC} value of all the carbons but showed the lowest DOM uptake on a surface-area basis. Therefore, increasing pH_{PZC} of the carbons toward more basic values was not the major factor responsible for the enhanced DOM uptake. It appears that the surface area in pores >1 nm played a more important role among the steam-treated carbons having basic pH_{PZC} values.

In general, the decrease in the DOM uptake with increasing carbon surface acidity (i.e., with acidic pH_{PZC} values) is primarily due to presence of ionized functional groups (e.g., carboxylic type) which carry a negative charge (at the pH values of isotherms) and repels DOM molecules approaching the activated carbon surface. The number of strong acidic functionalities on the carbon surface is significantly decreased during the steam treatment (800°C) due to the decomposition of such functionalities. The results obtained in this study indicate that when most of these acidic groups are removed and carbon pH_{PZC} increases toward basic values, further increases in the pH_{PZC} of the carbons (e.g., from pH_{PZC} of 8 to 10) may not significantly impact the DOM uptake. However, it is also noted that on a surface-area basis (i) the virgin HD (pH_{pzc}=5.3) showed a similar uptake as HD-S21 (pH_{pzc}=8.5), and (ii) SAC (pH_{pzc}=9.2) showed higher uptake than HD-S45 (pH_{pzc}=10.2). Considering the fact that comparison of surface-area normalized isotherms is based on the assumption that *all* pores >1 nm are accessible to all DOM components, and this approach does not consider any change in carbon *pore structure* as a result of steam treatment, the surface-area normalized isotherm may not be sensitive enough to demonstrate the impact of changes in carbon surface chemistry on the DOM uptake. In addition, there may be some additional carbon characteristics (pore structure or surface chemistry) that are not fully depicted by pore size distribution and pH_{PZC} measurements in terms of DOM uptake. Despite difficulties for resolving such subtle differences when dealing with heterogeneous materials like activated carbons and DOM, and limited techniques available for their quantitative characterization, the isotherm results clearly showed that steam treatment resulted in enhanced DOM uptake.

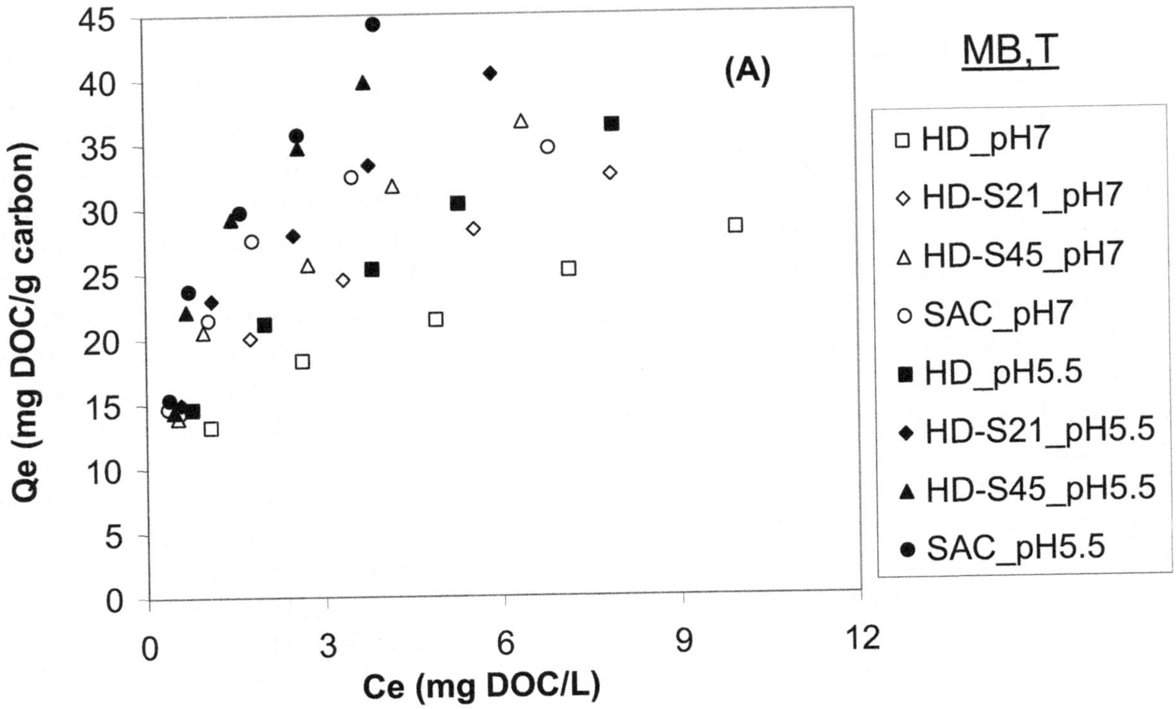

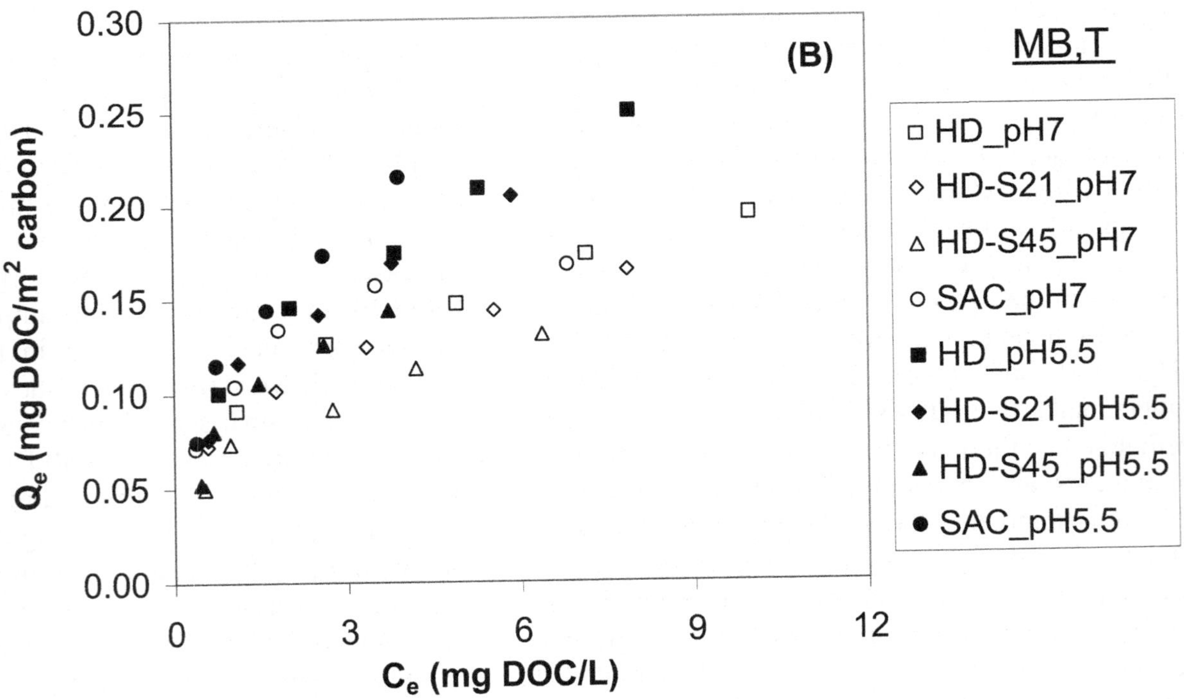

Figure 3.7 Adsorption isotherms of Myrtle Beach treated water DOM by virgin and steam treated carbons at pH 7 and pH 5.5: (A) mass basis, (B) surface-area basis

For each carbon, the DOM uptake at pH 5.5 was higher than that the uptake at pH 7. As discussed for carbon fibers, these results are consistent with the reported effect of pH in the literature.

Myrtle Beach raw water

The mass and surface-area basis isotherms of Myrtle Beach raw water DOM at pH 5.5 and pH 7 are shown in Figure 3.8. For mass-basis isotherms, the uptake of the carbons at pH 7 followed the order of SAC ≈ HD-S45 > HD-S21 ≈ HD. At pH 5.5, the order of uptake was HD-S45 ≥ SAC ≈ HD-S21 > HD. The raw water DOM uptake was lower than treated water uptake. Raw water contains larger molecular weight DOM components that are not present in the treated water (Figure 3.1). Therefore, at the same TOC concentration, the higher DOM uptake observed for the treated water was attributed to the presence of a higher amount of low molecular weight DOM components that can access a larger portion of the activated carbon surface area for adsorption.

HD-S21 carbon showed a higher DOM uptake as compared to the virgin HD in the treated water (Figure 3.7), but it did not in the raw water (Figure 3.8). A similar DOM uptake in the raw water suggests that the change in the pore structure of the HD-S21 was not large enough or not in the right size range to cause a difference in the adsorption of large DOM components in the raw water. However, the same change was enough to make a difference in the adsorption of low molecular weight components in the treated water DOM.

On a surface-area basis, at pH of 7 HD and SAC showed higher DOM uptakes than HD-S21 and HD-S45, despite significant differences in their pH_{PZC} values (Figure 3.8). On the other hand, at pH of 5.5, the trends were less apparent. As observed with the treated water, surface-area normalized isotherms did not correlate with the pH_{PZC} of the carbons. Overall, as mentioned before, the surface-area basis isotherm results should be interpreted with some degree of caution here; because not all DOM molecules in the raw water may use all the surface area in pores > 1 nm.

The decrease in pH increased the adsorption of both raw and treated water DOM in all carbons. The difference between the DOM uptake in raw and treated water by the same carbon was smaller at pH 5.5 than at pH 7. At pH 5.5 and C_e of 3 mg TOC/L, the DOM uptake of HD, HD-S21, HD-S45 and SAC in the treated water was about 23%, 27%, 5%, and 37%, respectively, higher than in the raw water. At pH 7, the DOM uptake of HD, HD-S21, HD-S45 and SAC in the treated water was 39%, 93%, 59% and 90% higher than in the raw water. The observed difference between the raw and treated water DOM uptakes of carbons with decreasing pH is probably due to more effective coiling of the large DOM components in the raw water.

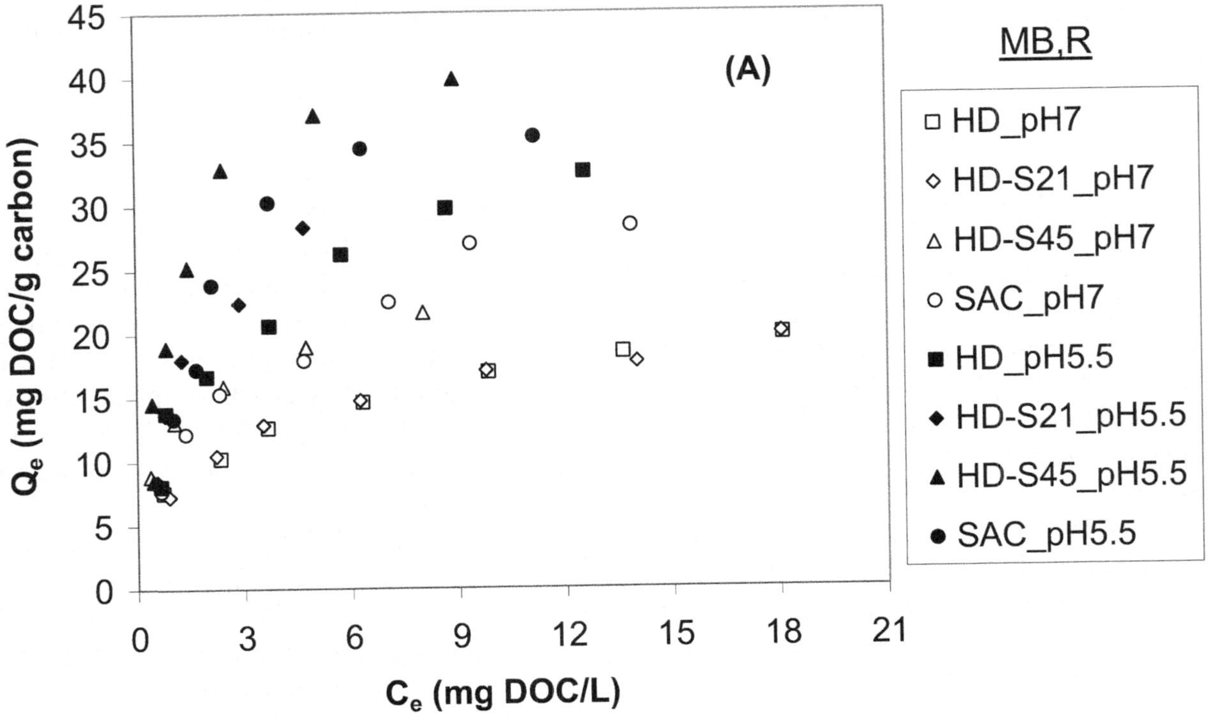

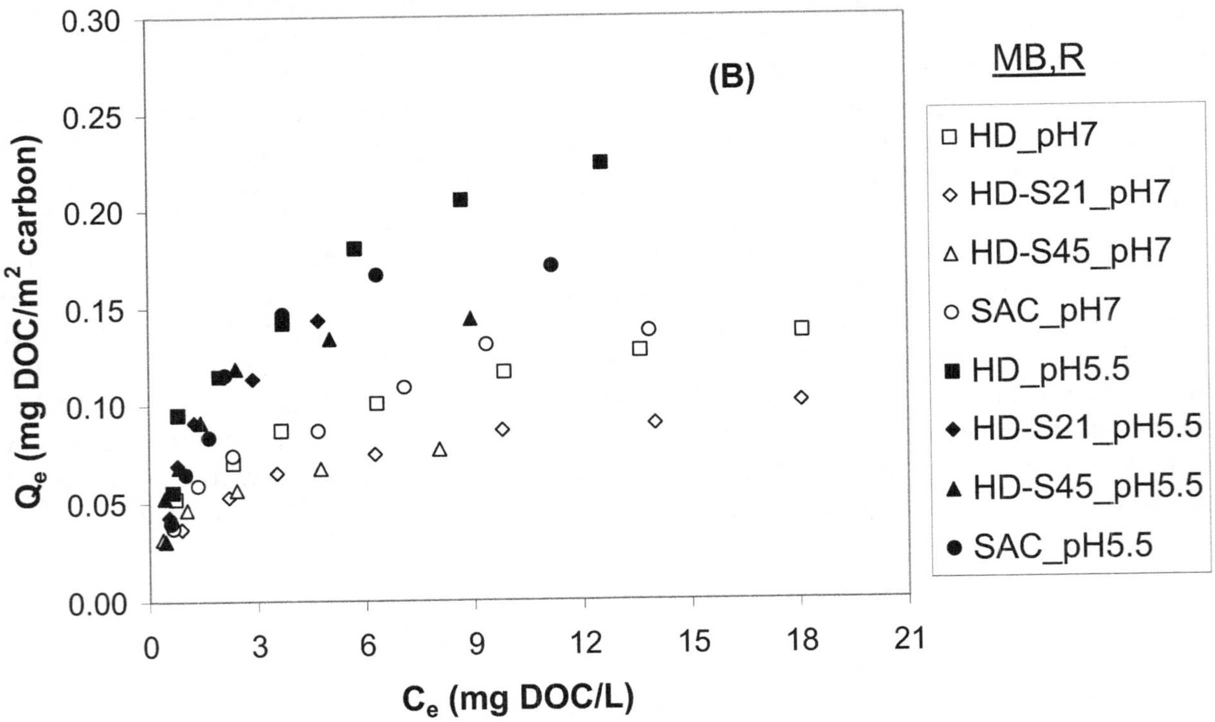

Figure 3.8 Adsorption isotherms of Myrtle Beach raw water DOM by virgin and steam treated carbons at pH 7 and pH 5.5: (A) mass basis, (B) surface-area basis

Spartanburg treated water

Mass and surface-area basis DOM isotherm results are shown in Figure 3.9. On a mass basis, the DOM uptake of the carbons at pH 7 followed, in general, the order of SAC $\geq$ HD-S45 $\approx$ HD-S21 >> HD, while the order of DOM uptake at pH 5.5 was SAC > HD-S45 $\geq$ HD-S21 >> HD. It is evident that increase in both accessible surface area (i.e., pores larger than 1 nm) and pH_{PZC} of the carbons after steam treatment enhanced the DOM adsorption. However, the extent of benefit from the steam treatment appears to be related to DOM characteristics as well as stream treatment conditions. A comparison of Myrtle Beach and Spartanburg treated water DOM uptake trends by HD, HD-S21, and HD-S45 shows the role of DOM characteristics because these carbons have the same precursor (i.e., HD). The DOM uptake trends at pH 5.5 and 7 are HD-S45 $\geq$ HD-S21 >> HD and HD-S45 > HD-S21 >> HD for Spartanburg and Myrtle Beach treated waters, respectively. While the higher accessible pore volume and higher pH_{PZC} as a result of steam treatment resulted in higher uptake of the Myrtle Beach treated water DOM by HD-S45 than HD-S21, the difference between the Spartanburg treated water DOM adsorption by HD-S45 and HD-S21 was negligible (Figure 3.9). For both waters, HD-S21 exhibited significantly higher DOM adsorption than the virgin HD. Opening carbon pores by steam treatment enhanced the DOM uptake, however extended steam treatment period and additional pore enlargement did not result in extra DOM adsorption for the Spartanburg treated water DOM. Myrtle Beach and Spartanburg treated DOM had similar $SUVA_{254}$ and MW_w values. It appears that there were differences (e.g., the *size* and structure) between these two DOM mixtures that were not fully resolved by $SUVA_{254}$ and MW_w measurements, causing differences in their adsorption by porous activated carbons.

In terms of the role of carbon characteristics and steam-treatment protocol, it would be expected that HD-S45 shows the highest DOM uptake among all GACs tested because it has the highest pore volumes in pores > 1nm and pH_{PZC}. However, it did not always show the highest DOM uptake. HD-S45 and SAC were prepared using the same lignite based precursor (i.e., Norit's HD) under steam-treatment, but in two different laboratories using different carbon batches and steam-treatment protocols. Although SAC has lower surface area in pores > 1 nm and one unit lower pH_{PZC} than HD-S45, it showed similar and sometimes higher DOM uptakes than HD-S45. Therefore, it appears that differences in steam treatment protocols and/or raw materials may result in some variations in the characteristics of the final product (e.g., carbon pore structure and/or surface chemistry) that are not completely resolved using the conventional techniques available (e.g., surface area and pore size distribution from nitrogen gas adsorption) for carbon characterization. Such differences may also impact the extent of DOM adsorption.

These findings show that GACs with wider pore size distributions and higher pH_{PZC} can be produced by extended steam treatment periods during carbon preparation. These steam-treated GACs are expected, in general, to show higher DOM uptake as compared to their precursors; however, the increase in DOM uptake is not always proportional to the extent of steam treatment. It also appears to be related to the physicochemical characteristics of DOM. The increase in the DOM uptake was not found to be correlated with simple DOM characterization parameters, such as $SUVA_{254}$ and MW_w determined through SEC used in this study. Finding the optimum steam-treated GAC appear to be water specific and will require experimental testing.

To examine the effect of carbon surface chemistry, the trends for the surface-area normalized isotherms were examined. In general, for both pH 5.5 and pH 7, the surface-area basis DOM uptake of Spartanburg treated water followed the order of SAC $\approx$ HD $\approx$ HD-S21 >

HD-S45 (Figure 3.9). The order for the Myrtle Beach treated water was SAC > HD ≈ HD-S21 > HD-S45. These orders did not correlate with the pH_{PZC} values of the carbons; HD-S45 had the highest pH_{PZC} value of all carbons but showed the lowest DOM uptake on a surface-area basis. Therefore, this analysis further confirms that increasing pH_{PZC} of the carbons toward more basic values was not the major factor responsible for the enhanced DOM uptake.

For each carbon, the DOM uptake at pH 5.5 was higher than that at pH 7 for both Spartanburg and Myrtle Beach treated water DOM. As discussed for carbon fibers, these results are consistent with the literature, and the observed increase can be attributed to a combination of various factors, as previously listed, that become important with decreasing pH.

Spartanburg raw water

Only mass basis isotherms will be presented and discussed. The mass basis isotherms of Spartanburg raw water DOM are shown in Figure 3.10. The DOM uptakes of the carbons at pH 7 followed the order of SAC > HD-S45 ≈ HD-S21 > HD and SAC ≈ HD-S45 ≈ HD-S21 > HD at pH 5.5. Again, steam treatment enhanced the DOM uptake from the raw water. As for the treated waters, the trends indicate that longer steam treatment periods (i.e., HD-S45 vs HD-S21) had less positive impact on the Spartanburg raw water than on the Myrtle Beach raw water DOM adsorption. This may be due to the larger size of DOM molecules in Myrtle Beach raw water. Furthermore, a one and one half unit decrease in pH increased the DOM adsorption for both waters; however, the pH effect was greater for Myrtle Beach raw water DOM. This suggests better coiling of DOM molecules and/or their interactions with the carbon surface at lower pH in Myrtle Beach than Spartanburg water, which may relate to the pK_a distribution of these DOM. As for the Myrtle Beach water, the DOM uptakes by activated carbons were lower from the raw water than from the treated water for Spartanburg samples.

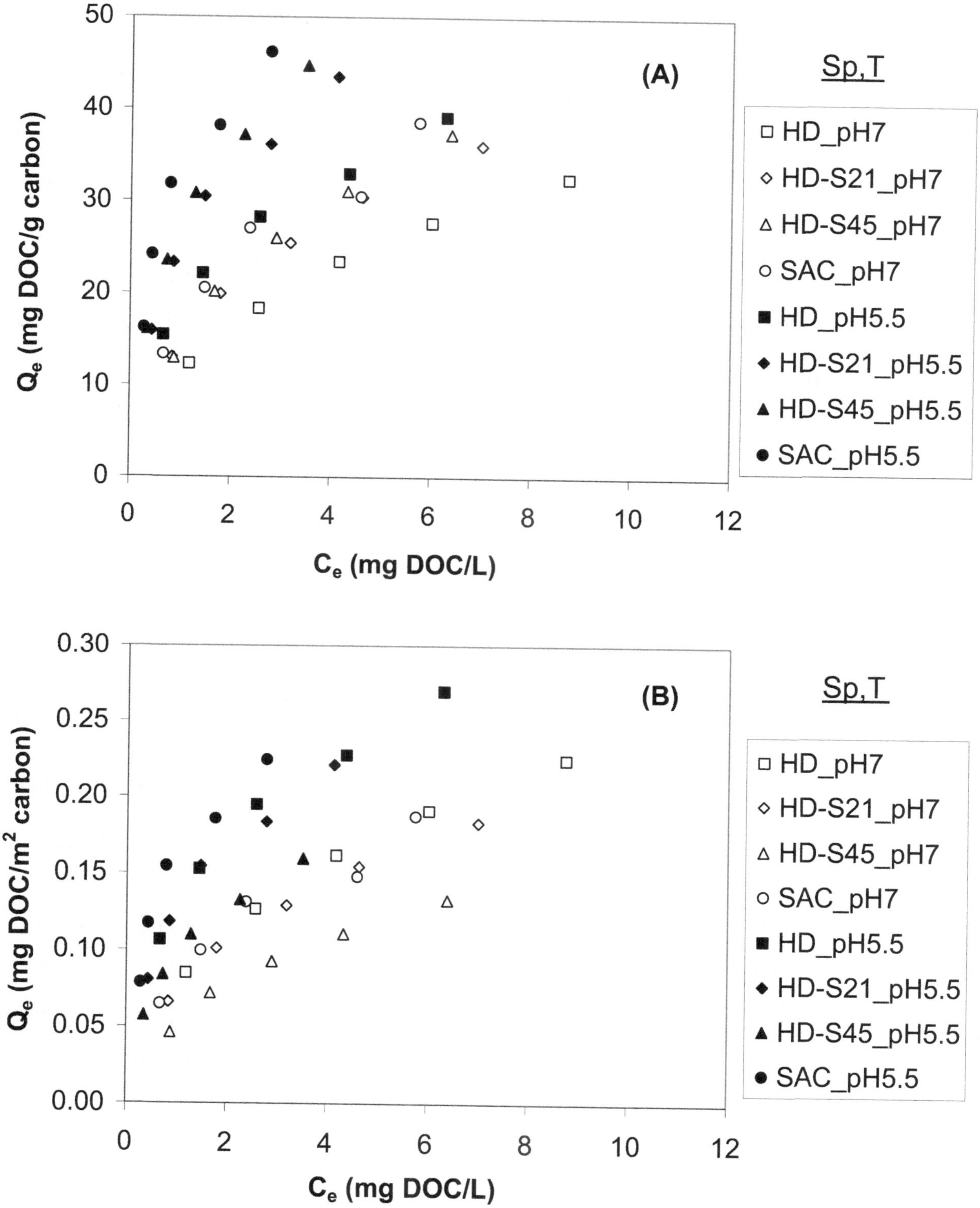

Figure 3.9 Adsorption isotherms of Spartanburg treated water DOM by virgin and steam treated carbons at pH 7 and pH 5.5: (A) mass basis, (B) surface-area basis

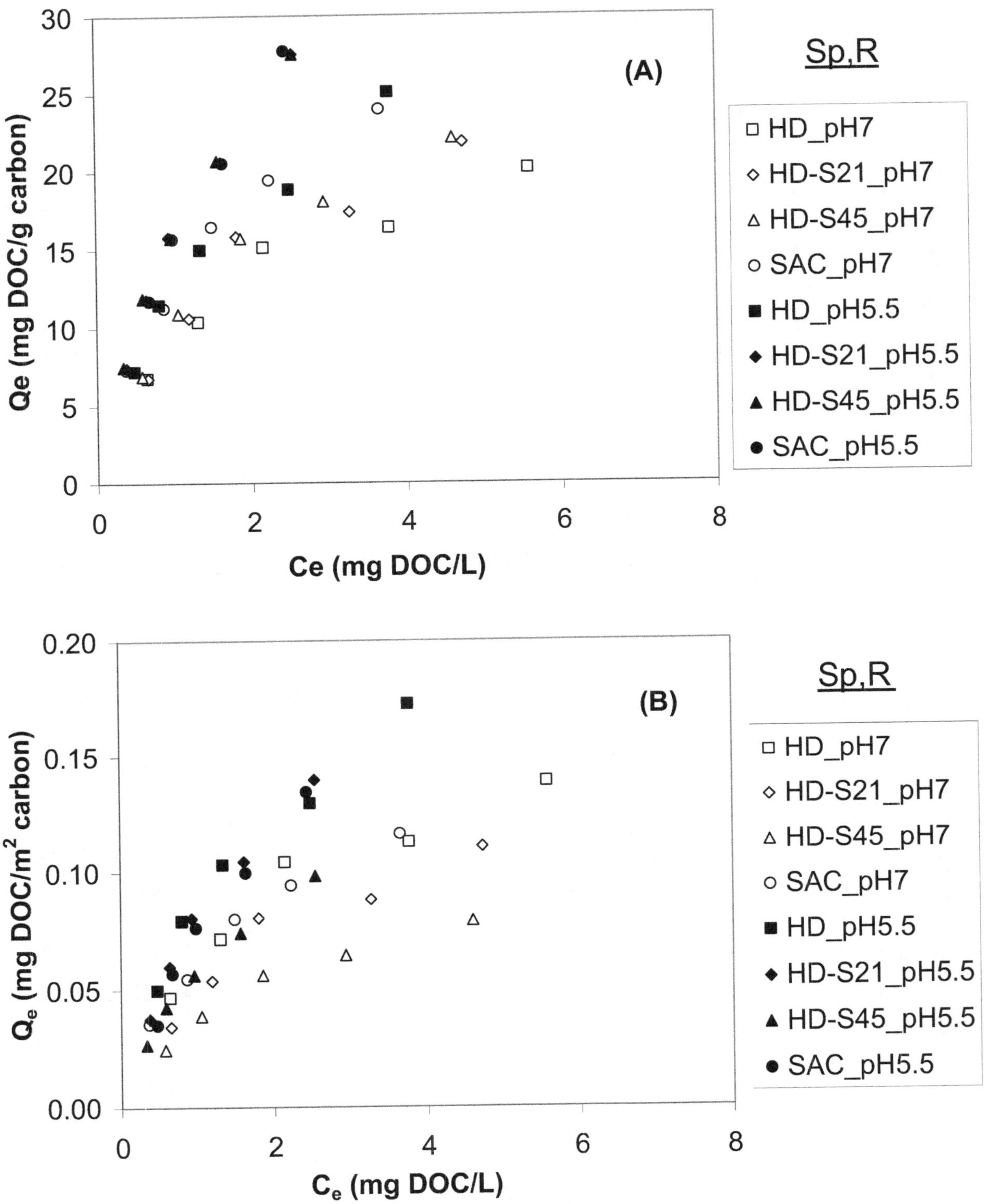

Figure 3.10 Adsorption isotherms of Spartanburg raw water DOM by virgin and steam treated carbons at pH 7 and pH 5.5: (A) mass basis, (B) surface-area basis

RSSCT Results

The RSSCT results of the virgin and steam-treated carbons will be separately presented and discussed for Myrtle Beach and Spartanburg waters. The majority of RSSCTs were performed using the treated water DOM because of its practical relevance. However, for comparison purposes, a few raw water RSSCTs were also conducted using the virgin carbon.

The RSSCT breakthrough curves are presented as a function of both bed volume of water treated and large scale filter bed operation time. The breakpoint for a compound is defined as its maximum acceptable concentration at the effluent of a column. Since there is no regulatory standard for DOM, breakthrough profiles were obtained approximately up to 70% of influent concentration in this study. To determine the amount of THM and HAA_9 that can be formed in the column effluents, periodically collected breakthrough samples, approximately 10 points per experiment, were chlorinated according to UFC protocol as described in the Materials and Methods Section. The concentrations of THM and HAA_9 were also plotted as a function of bed volume of water treated and large scale filter bed operation time. To gain some insight into the nature of DOM captured in the column, batch resin fractionation experiments were performed on the periodically collected column effluent samples. The results are presented as the percent distribution of hydrophobic/hydrophilic DOM components as a function of bed volume of water treated and large scale filter bed operation time.

Virgin carbon results

Myrtle Beach water. The TOC and UV_{254} breakthrough profiles of both Myrtle Beach raw and treated water DOM were obtained for the virgin carbon at different pH values (Figure 3.11). The breakthrough patterns for TOC and UV_{254} removal were generally consistent with each other. For the raw water, a rapid increase of TOC and UV_{254} values was observed in the breakthrough profile, which indicated that some DOM components quickly eluted from the column. Apparently, these DOM components in the raw water were not amenable for adsorption. The % HPO profile showed that the DOM components initially eluting from the column were highly hydrophobic in character, 60-80% HPO, which was significantly higher than the 54% hydrophobicity in the influent raw water (Figure 3.12A). The $SUVA_{254}$ of the column effluents showed that the initial DOM components eluting from the column had also high $SUVA_{254}$ values (Figure 3.12B). In general, MW_w increases with increasing $SUVA_{254}$ of DOM. Although molecular weight of these DOM components were not measurable by SEC (equipped with a UV detector) due to their very low concentrations, the overall results suggest that the DOM components initially eluting from the column were large molecular weight hydrophobic and aromatic DOM components, which were excluded from carbon pores due to size exclusion effects.

The THM and HAA_9 formation in the column effluents as a function of breakthrough profiles is shown in Figure 3.13. Consistent with the TOC and UV_{254} profiles, the amount of THM and HAA_9 formation in the column effluents rapidly increased for the raw water. The plots of THM and HAA_9 normalized by TOC as a function of bed volume showed that the DOM components that initially eluted from the column had very high DBP formation reactivity (Figure 3.14). However, since the TOC concentrations were low, their overall THM and HAA_9 formation were still low. Strong correlations were observed between THM or HAA_9 and UV_{254}, indicating that the UV absorbing components play a major role in the DBP formation (Figure 3.15).

For the treated water, a much longer breakthrough profile was obtained without a rapid initial increase at the same influent concentration of 3 mg TOC/L (Figure 3.11). The TOC concentration at the column effluent remained around 0.2 mg/L, approximately equal to the minimum reporting level of the measurements, until 2800 bed volumes at pH 6.5. This showed that there was no significant amount of nonadsorbing DOM components in the treated water RO isolate. The % HPO and $SUVA_{254}$ profiles did not show the presence of highly hydrophobic and high $SUVA_{254}$ DOM components eluting from the column, especially at the beginning of the column run as observed in the raw water. All these DOM components in the raw water were removed during the conventional treatment processes in the drinking water treatment plant, and the DOM components remaining in the treated water were adsorbable by the virgin carbon.

Slowly increasing trends of the breakthrough profiles were observed after the effluents reaching around 60% of influent concentration. Such tailing of the breakthrough has been observed in full-scale applications (Roberts and Summers, 1982). This was attributed to biodegradation of organic matter and slow sorption kinetics of DOM. Biodegradation is unlikely to be a major factor in this study, due to the relatively short period (<15 days) of the experiments and prefiltration of stock solution using 0.45-μm filters. Slow sorption kinetics of DOM molecules in carbon pores is probably the main reason for the observed tailing behavior of breakthrough profiles. The %HPO profile for the experiment at pH 6.5 and 5.5 showed a gradually increasing pattern consistent with the TOC and UV_{254} breakthrough profiles (Figure 3.12). When approximately 60% of the breakthrough was reached for pH 6.5 and pH 5.5, %HPO started to level off around 40% which was about the %HPO of the influent water. The $SUVA_{254}$ breakthrough profiles also gradually approached to the $SUVA_{254}$ of the influent water with increasing bed volume (Figure 3.12).

The enhanced DOM adsorption with decreasing pH observed during the isotherm experiments was also observed during the column experiments. At 70% of TOC breakthrough, the volume of water treated at pH 5.5 was nearly 2760 bed volumes (32%) higher than that was obtained at pH 6.5 (Figure 3.11A). The nearly parallel shifts of the breakthrough patterns with decreasing pH indicated that although the DOM uptake of the GAC increased, there was no significant impact on the adsorption kinetics. Overall, maintaining a lower pH after conventional treatment processes, such as those used for enhanced coagulation, will have a positive impact on the carbon utilization in GAC filters. Finding the optimum pH from a perspective of treatment cost requires a combined evaluation of coagulation/flocculation with GAC adsorption (Hooper et al., 1996).

The formation of THM and HAA_9 in the column effluents for the Myrtle Beach treated water is shown in Figure 3.13. The experimental results have indicated that chloroform was the major THM species, and dichloroacetic acid and trichloracetic acid were the major HAA_9 species. The formation of bromochloro DBPs was small, most of the time less than 5 μg/L, since the bromide level in the influent was below the detection limit. No mono-, di- and tribromoacetic acid formation was observed at the column effluents. The benefit of one unit decrease in pH on the TOC removal was also reflected on the THM and HAA_9 formation. Strong correlations were obtained between the THM or HAA_9 and UV_{254} in column effluent samples (Figure 3.15).

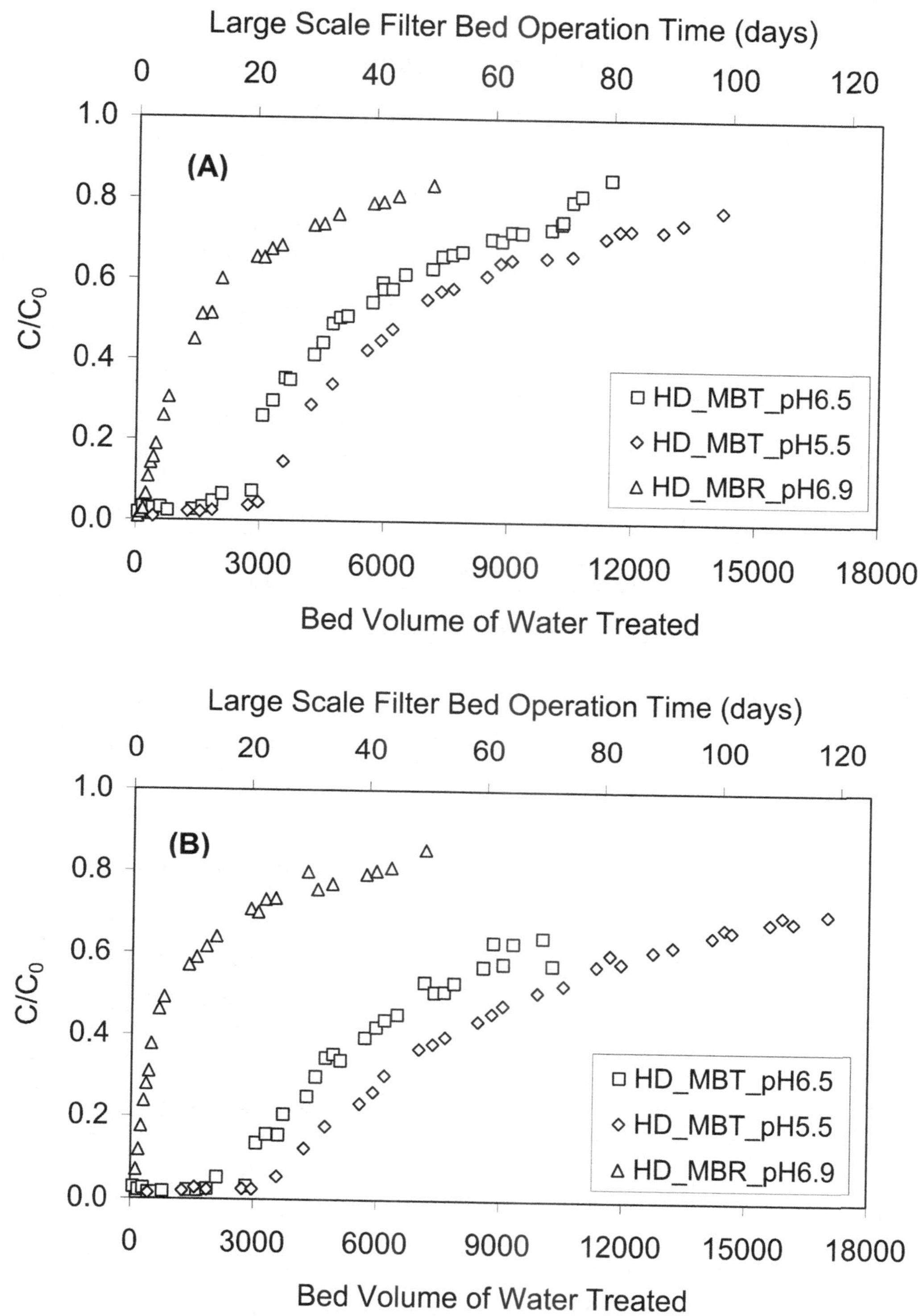

Figure 3.11 Myrtle Beach DOM breakthrough profiles, expressed as (A) TOC and (B) UV_{254}, for the virgin carbon. A constant C_o of 3 mg TOC/L was used in all experiments. MBT: Myrtle Beach treated water and MBR: Myrtle Beach raw water

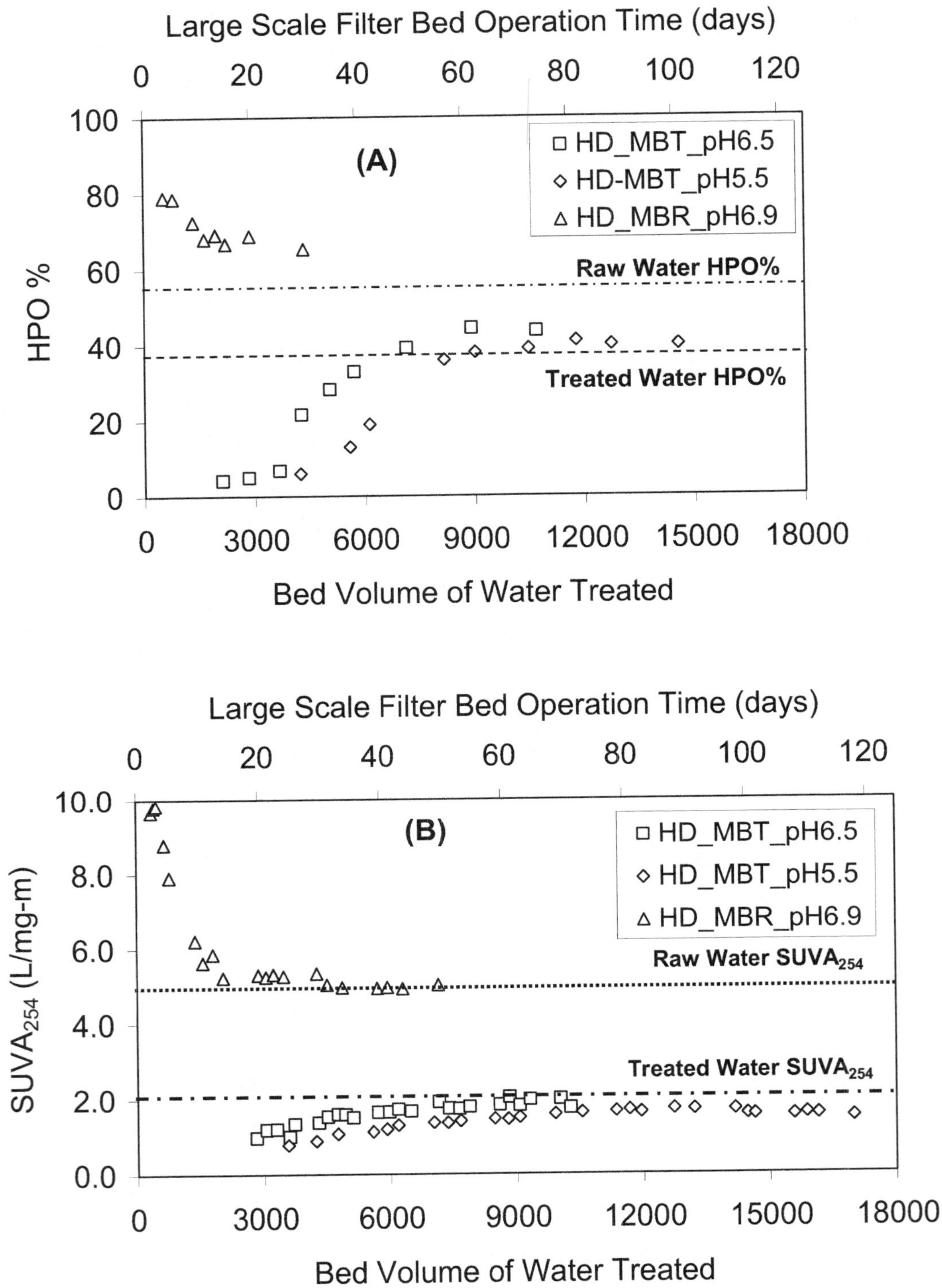

Figure 3.12 Myrtle Beach raw and treated water DOM (A) %HPO and (B) $SUVA_{254}$ breakthrough profiles for the virgin carbon. The dashed lines show the %HPO and $SUVA_{254}$ of the raw and treated waters

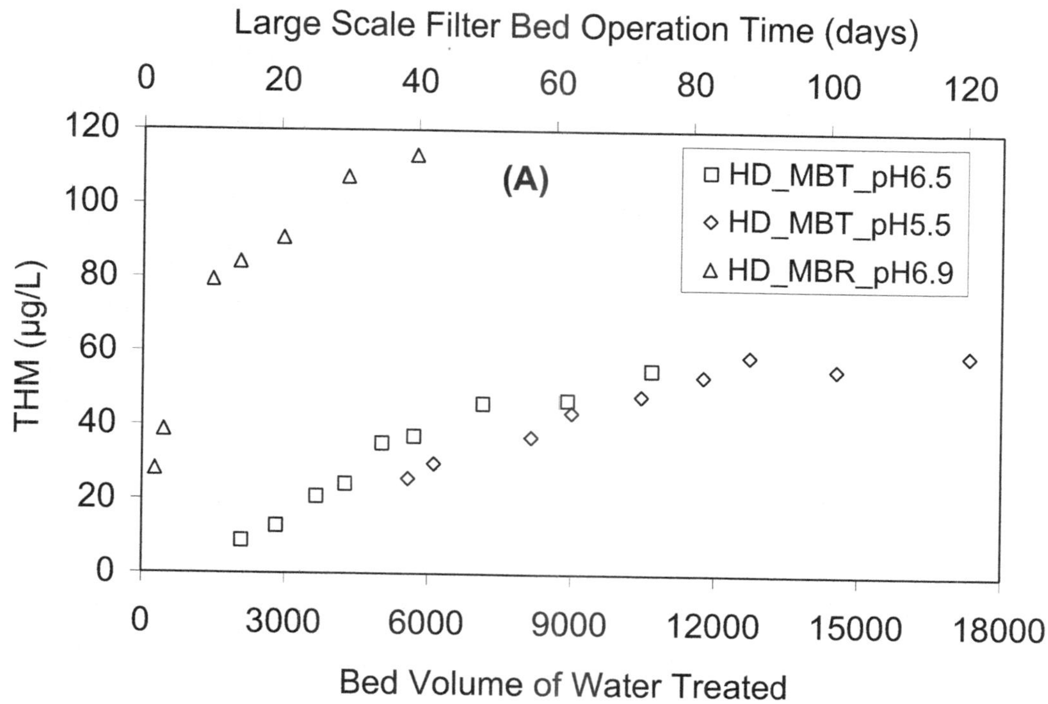

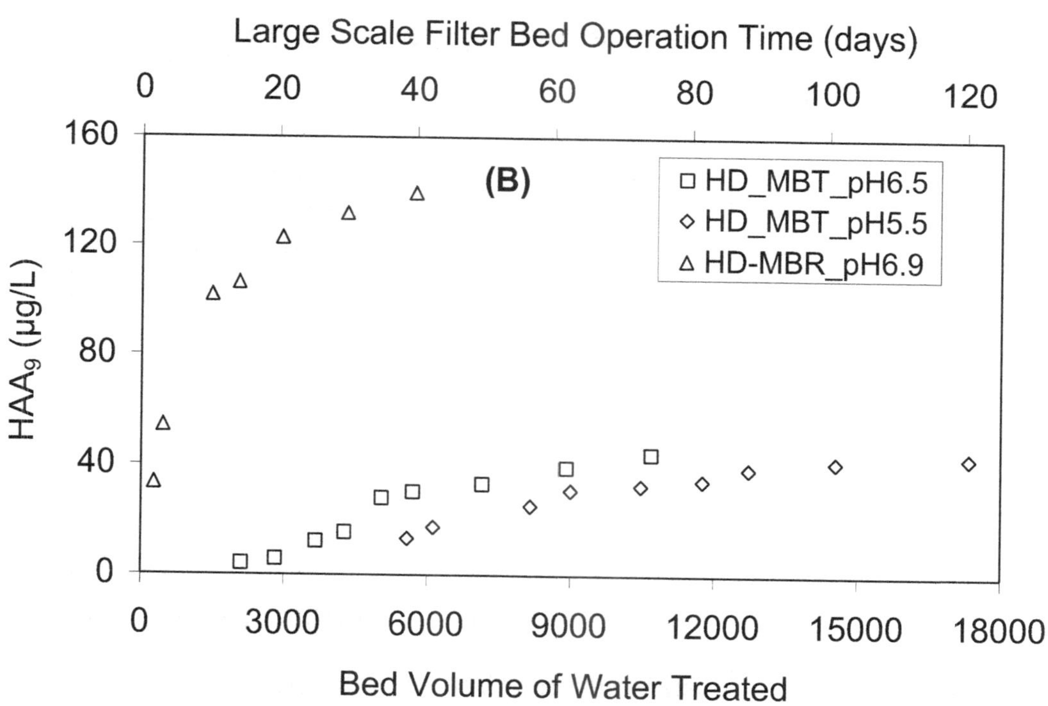

Figure 3.13 Myrtle Beach raw and treated water (A) THM and (B) HAA_9 formation breakthrough profiles for the virgin carbon

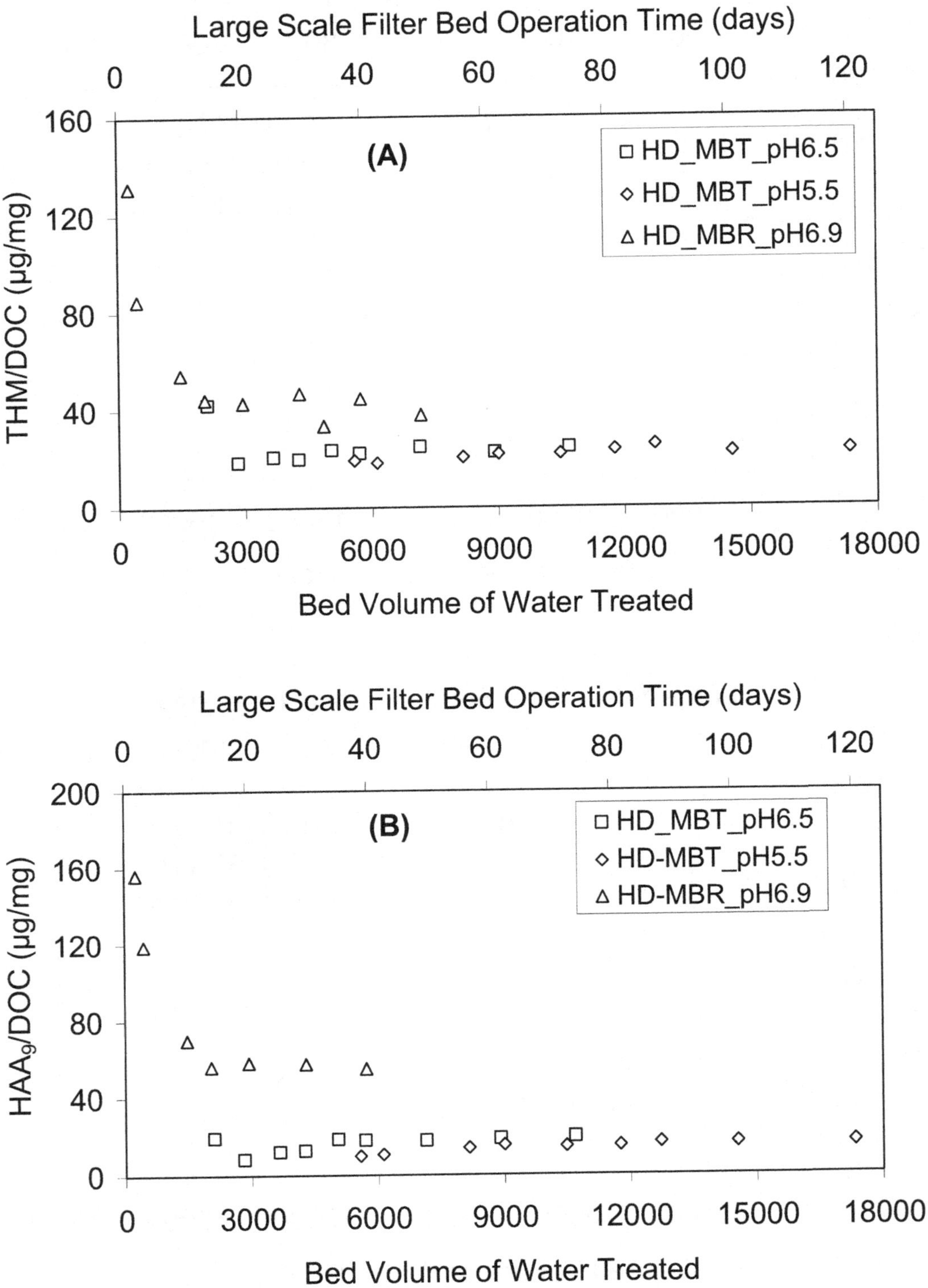

Figure 3.14 (A) THM and (B) HAA_9 reactivity of Myrtle Beach raw and treated water DOM eluting from the virgin carbon column

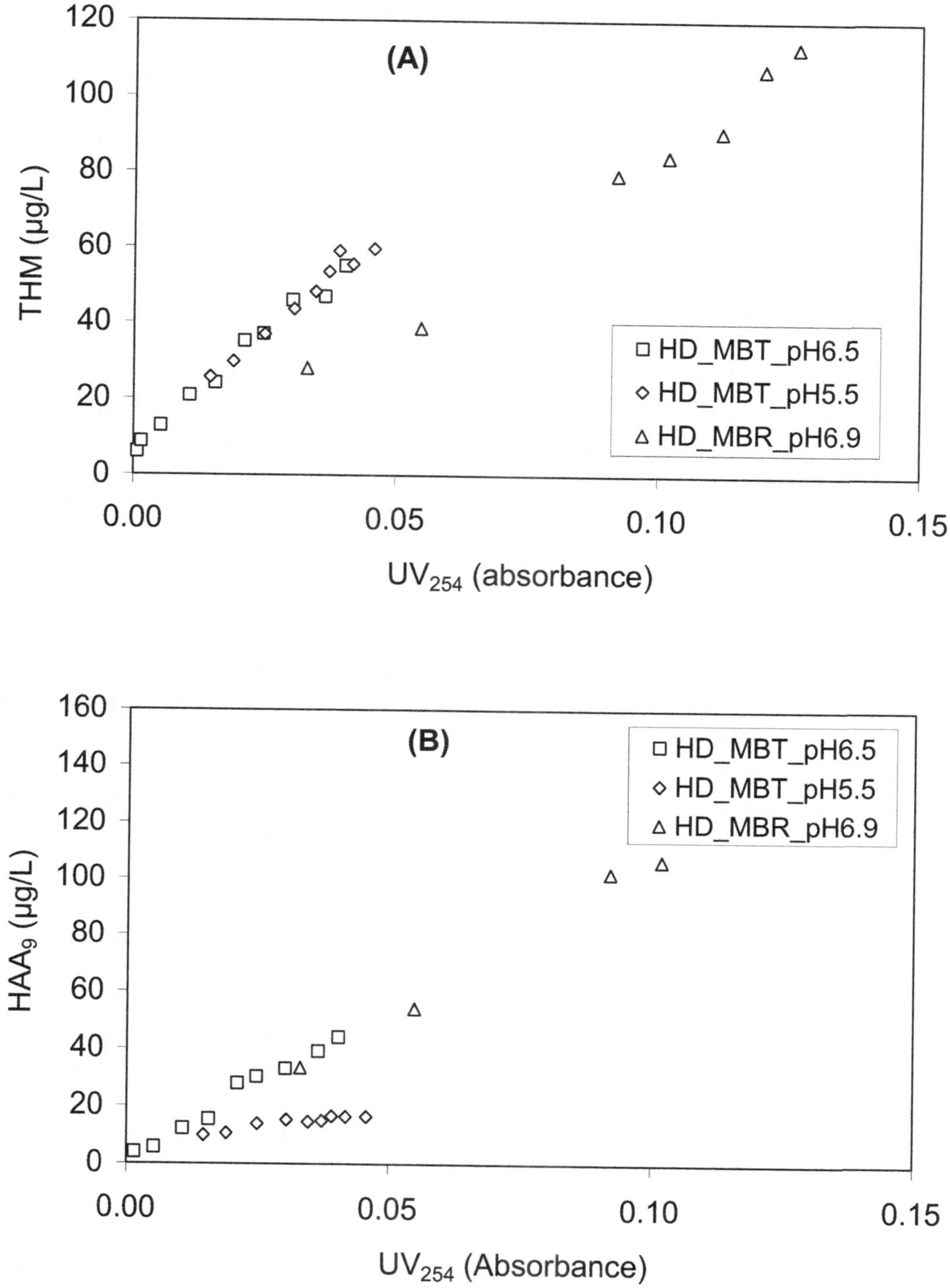

Figure 3.15 THM vs. UV_{254} and HAA_9 vs. UV_{254} of Myrtle Beach raw and treated water DOM eluting from the virgin carbon column

Spartanburg water. The breakthrough profiles for the Spartanburg water are shown from Figures 3.16 to 3.18. The results were, in general, consistent with those obtained from the Myrtle Beach raw water. A rapid increase in the breakthrough profile of the raw water was observed, and the DOM components eluting initially from the column were high $SUVA_{254}$ components. In contrast to the Myrtle Beach DOM, the HPO% profile indicated that these high $SUVA_{254}$ DOM components in Spartanburg raw water showed hydrophilic character. This indicates a significant difference in the DOM characteristics between Myrtle Beach and Spartanburg water. However, as with Myrtle Beach raw water, the DOM components initially eluting from the column had a high degree of THM and HAA_9 reactivity (Figure C.1).

Spartanburg treated water results were also in agreement with those obtained from Myrtle Beach treated water. However, it was noted that the one unit decrease in pH had no impact on the DOM removal nor on THM and HAA_9 formation (Figures 3.16 and 3.18). This was in disagreement with the DOM removal observed in the batch isotherms (Figure 3.9). At 3 mg/L equilibrium concentration, isotherm results showed a higher uptake at pH 5.5 than pH 6.5. This difference is probably due to different adsorption pattern between column and isotherm experiments. Further discussion on this issue will be presented after the steam-treated carbon results. The %HPO, THM and HAA_9 experiments results were generally consistent with the TOC breakthrough profiles.

Steam-treated carbon results

The RSSCTs of Myrtle Beach and Spartanburg raw and treated water DOM were performed for the steam-treated GACs at both pH 5.5 and pH 6.5. The results of each water will be discussed separately. Since GACs are commonly used after conventional C/F/S processes in practice, limited experiments were performed with the raw waters.

Myrtle Beach raw water. The breakthrough profile was obtained for one steam-treated carbon (HD-S21) at pH 6.5 (Figure 3.19). The result for the virgin carbon is also added in this figure for comparison purpose. Mild steam-treatment (i.e., HD-S21) had a slight impact on the removal of Myrtle Beach raw water DOM, as observed in the isotherm experiments (Figure 3.8). As for the virgin carbon, a rapid breakthrough pattern was observed for the steam-treated carbon. The eluting DOM components were initially highly hydrophobic and high $SUVA_{254}$ DOM components (Figure C.3). The immediate elution of the raw water DOM from the HD-S21 column further confirmed the presence of some nonadsorbable (probably high molecular weight) DOM components in the water. The early portion of the breakthrough curves of HD and HD-S21 were nearly identical. After approximately 1320 bed volume, the breakthrough curve of HD carbon split from that of HD-S21 and displayed a faster breakthrough behavior. The higher surface area in pores >1 nm of HD-S21 than HD probably provided slightly better adsorption kinetics for some DOM components.

The THM and HAA_9 breakthrough profiles were consistent with those of TOC and UV_{254} (Figure 3.20). Interestingly, the results showed that steam treatment decreased the HAA_9 formation significantly, whereas no impact was observed on the THM formation. It appears that some additional removal of TOC (although not very significant in term of TOC mass removed per carbon mass) during the later stage of the breakthrough preferentially removed HAA_9 precursors in the raw water. The DOM components initially eluting from the column exhibited very high reactivities of THM and HAA_9 formation, as previously documented for the virgin carbon sample (Figure C.4). In addition, strong correlations were observed between THM or HAA_9 formations and UV_{254} (Figure C.5).

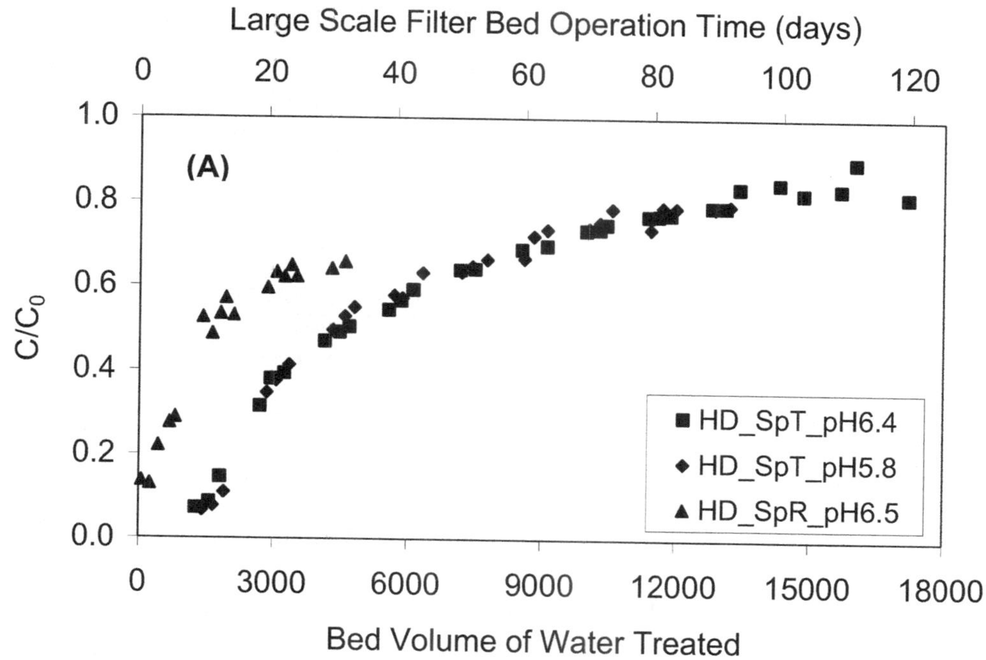

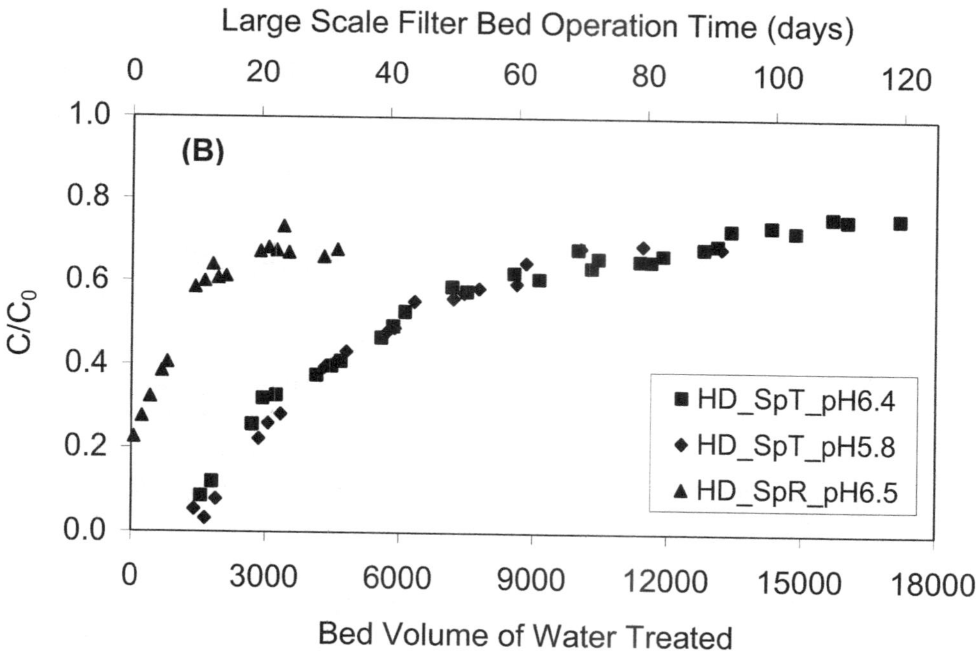

Figure 3.16 Spartanburg DOM breakthrough profiles, as (A) TOC and (B) UV_{254}, for the virgin carbon. SpT: Spartanburg treated water, SpR: Spartanburg raw water.

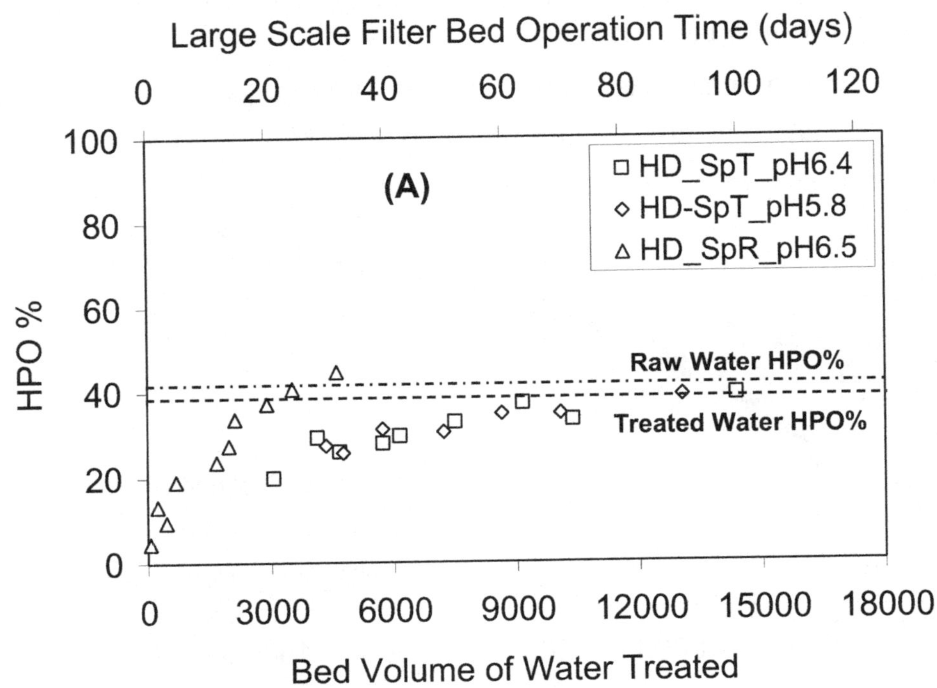

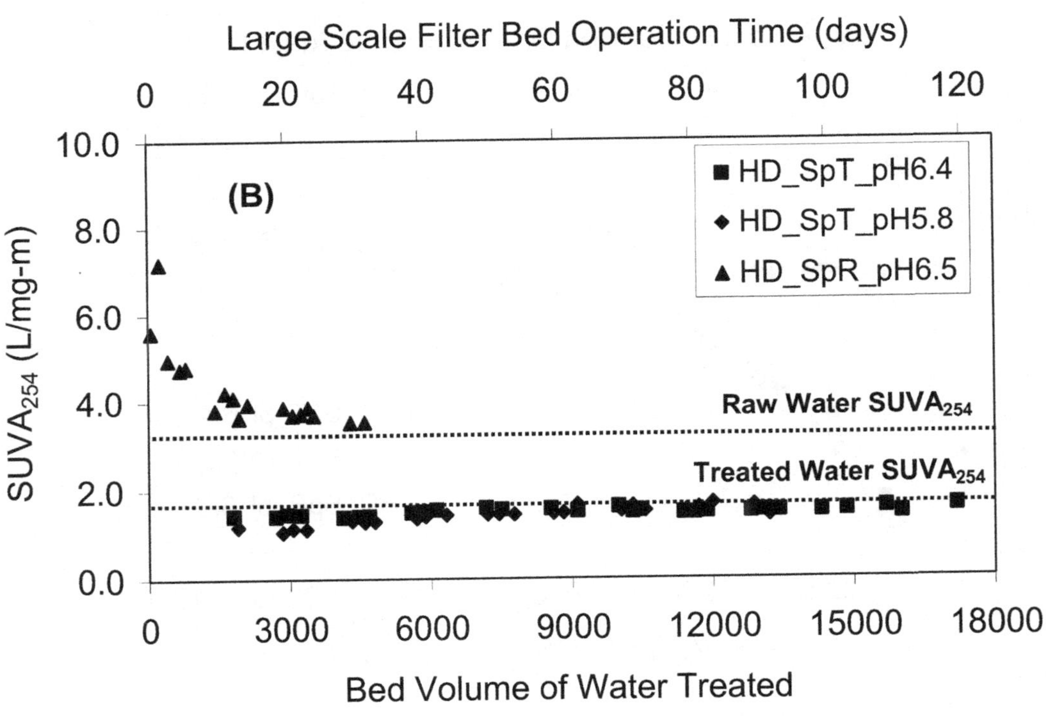

Figure 3.17 Spartanburg water (A) %HPO and (B) $SUVA_{254}$ breakthrough profiles for the virgin carbon

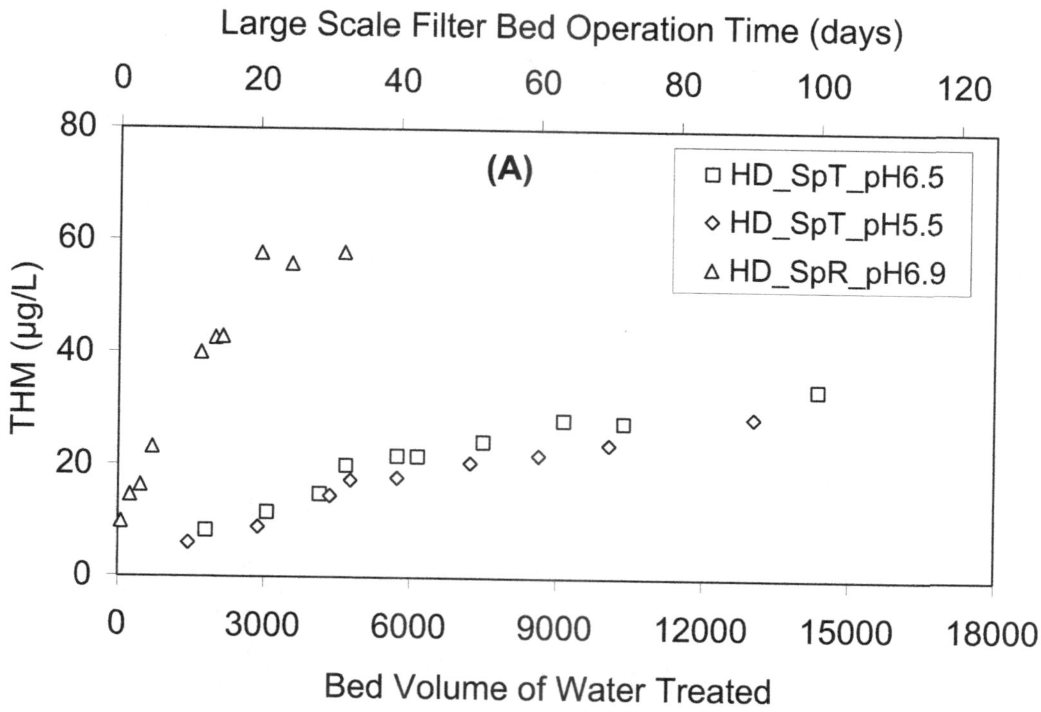

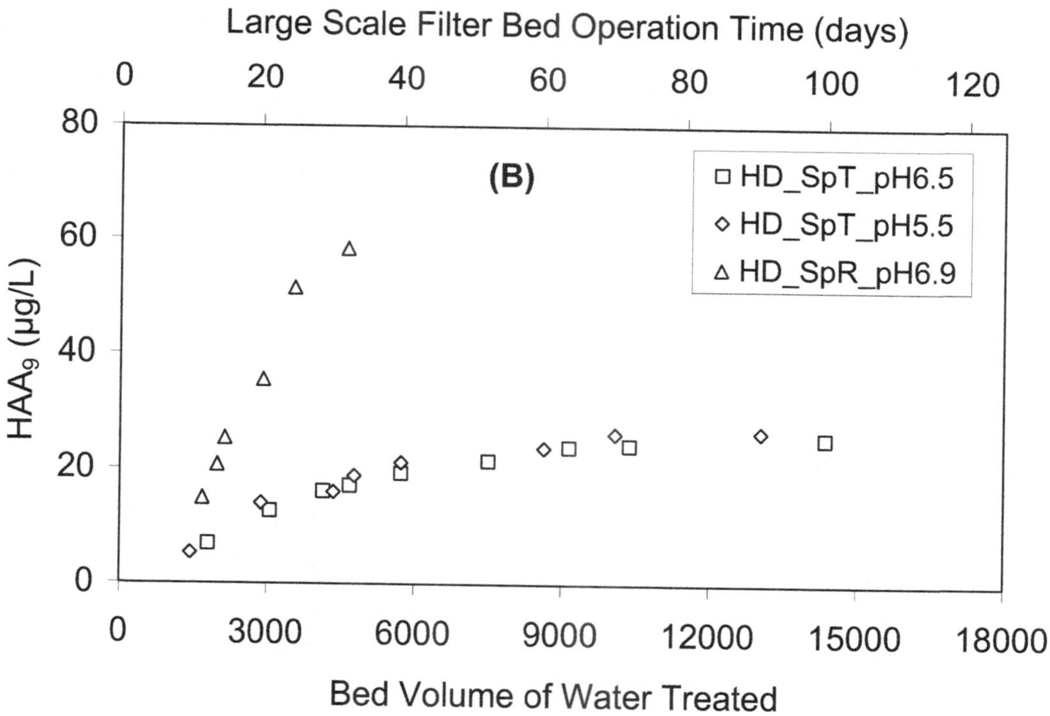

Figure 3.18 Spartanburg water DOM: (A) THM and (B) HAA_9 formation breakthrough profiles for the virgin carbon

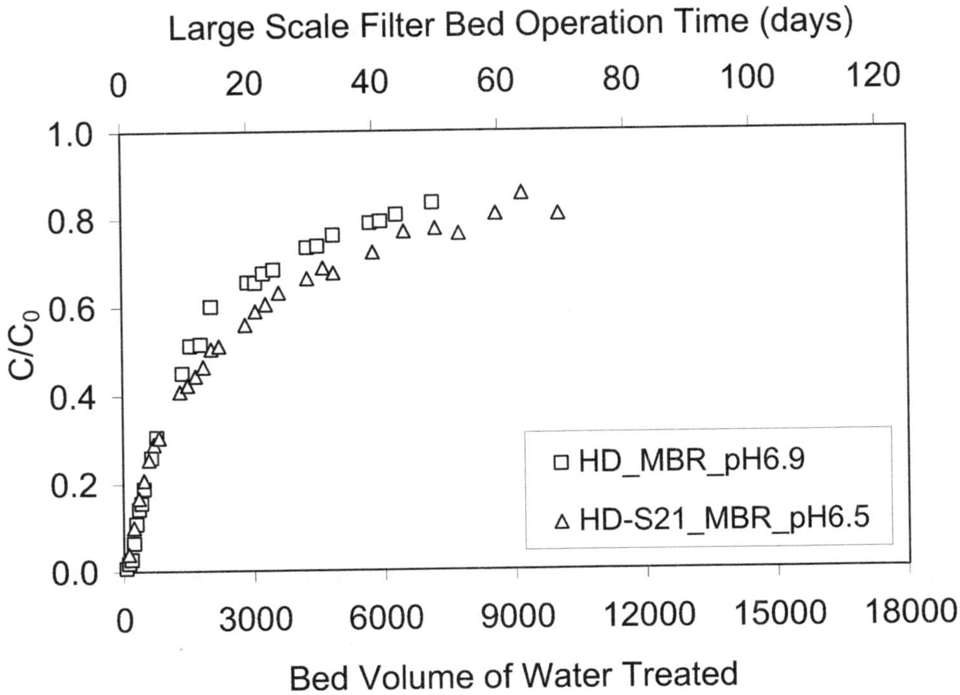

Figure 3.19 Myrtle Beach raw water DOM (TOC) breakthrough profile for virgin and steam-treated carbons

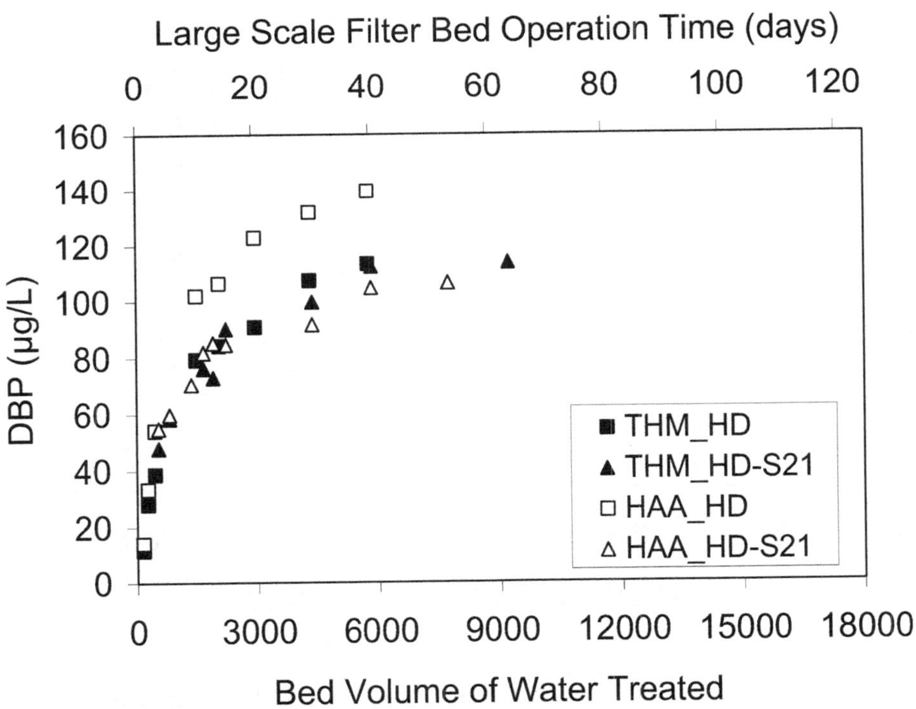

Figure 3.20 DBP formation of Myrtle Beach raw water DOM for virgin and steam treated carbons breakthrough profiles at pH 6.5

Myrtle Beach treated water. The RSSCT results for the Myrtle Beach treated water are presented with various plots from Figures 3.21 to 3.28. The RSSCT effluent TOC concentrations of the treated water DOM remained around the detection limit until 2880 bed volumes indicating that the majority of the DOM components in the treated water adsorbed on the virgin and steam-treated GACs (Figure 3.21). Steam-treated carbons exhibited significantly higher DOM uptake than the virgin HD. As observed for the isotherms, the extent of DOM removal was in the order of HD-S45>HD-S21>HD. The additional TOC removal resulted in a delayed increase in the %HPO and $SUVA_{254}$ breakthrough profiles as a function of bed volume of water treated (Figure 3.22). This indicates that DOM components with more hydrophobic character and having higher $SUVA_{254}$ values were removed by steam-treated GACs.

THM and HAA_9 results for breakthrough samples are provided in Figure 3.23. There are three important observations about these profiles. (1) Despite the enhanced TOC removal as a result of steam treatment, the impact on the THM formation was minimal, whereas a clear reduction was observed for HAA_9 formation. This indicates that the additional DOM removed by the steam-treated carbons mainly constituted the HAA_9 precursors in this water, (2) Despite a higher degree of TOC removal observed for HD-S45 as compared to HD-S21 GAC, the benefit in terms of THM and HAA_9 control was negligible. It appears that there is an maximum benefit that can be obtained from steam treatment by enlarging carbon pores to control the DBP formation, and (3) Although TOC removal is more feasible to monitor the performance of a GAC column, determination of THM and HAA_9 formation in the column effluent is necessary to judge the performance of a GAC with respect to compliance with the DBP regulations. Therefore, an RSSCT experiment should be combined with DBP determination in selecting activated carbons for DBP formation control. The THM and HAA_9 reactivity profiles, and the plots of THM and HAA_9 formation correlations with UV_{254} are provided in Figure C.6 and C.7. The results, in general, are in agreement with those observed from the virgin carbon.

Myrtle Beach treated water: pH Effect. The RSSCTs at pH 5.5 were also conducted for HD-S21 and HD-S45 to investigate the pH effect. The TOC breakthrough profiles were compared with that of the virgin GAC at both pHs (Figure 3.24). The UV_{254} breakthrough profiles were consistent with TOC profiles (Figure C.8). A one unit decrease in pH increased the volume of water treated up to 70% breakthrough by about 30% for both virgin and steam-treated GACs (Table 3.7). The HD-S45 at pH 5.5 treated almost twice the water volume (up to 70% breakthrough) as compared to the virgin HD at pH 6.5. Therefore, it appears that by using a GAC with larger pores (e.g., steam-treated) and decreasing pH about one unit, it is possible to double the treated water volume in a carbon filter for DOM removal. The %HPO and $SUVA_{254}$ breakthrough profiles for steam-treated carbons were slightly delayed at pH 5.5 as compared to pH 6.5 (Figure 3.25). As discussed before, DOM molecules become smaller due to the coiling of molecules and more hydrophobic with decreasing pH.

THM and HAA_9 results for breakthrough samples are provided in Figures 3.26 and 3.27. The results showed significant reduction in both THM and HAA_9 formation as a result of steam treatment and decreasing pH. Although no major impact of steam treatment was observed on THM formation at pH 6.5 for either HD-S21 or HD-S45 (despite enhanced TOC removal as seen in Figure 3.24), a significant improvement was observed at pH 5.5. It appears that the decrease in pH assisted in the removal of some THM precursors. The steam-treated GACs, both HD-S21 and HD-S45, showed lower HAA_9 formation at the column effluent as compared to the virgin carbon. Although steam treatment decreased the HAA_9 formation, the additional benefit of the pH effect was only observed for HD-S45. The reactivity and UV correlation plots were generally consistent with other samples (Figure C.9 and C.10).

It is evident from these results that performing RSSCTs combined with THM and HAA_9 formation measurements is necessary to assess the performance of GACs. Furthermore, steam-treated GACs or GACs with larger mesopore volumes and higher pH_{PZC} values can significantly increase the effectiveness of DBP formation control with adsorption. The results for the Myrtle Beach treated water clearly show that after 18,000 bed volumes and approximately 4+ months of operation, the steam-treated HD-S45 at pH 5.5 had effluent THM and HAA_9 formations well below the regulatory limits, 80 and 60 ppb, respectively.

Spartanburg treated water. The TOC and UV_{254} breakthrough profiles for the steam-treated carbons on Spartanburg treated water DOM at pH 6.5 and 5.5 are shown in Figure 3.28. The results indicated that although steam-treated GACs showed higher DOM removal than the virgin carbon, the advantage was much less compared to the Myrtle Beach treated water. In addition, the difference between DOM removal of the two steam-treated GACs, HD-S21 and HD-S45, was negligible. Enlargement of carbon pores had much less effect on the Spartanburg water DOM. This was consistent with the observations from isotherm experiments, which was attributed to the more hydrophilic nature of Spartanburg water DOM.

Another important observation for the Spartanburg treated water results was the absence of a pH effect on the DOM removal. There was no significantly additional DOM removal for all virgin and steam-treated GACs when pH was lowered from 6.5 to 5.5. Similarly no pH effect was observed on THM and HAA_9 removal with HD-S21 (Figure 3.29). Overall, one unit decrease in pH did not have a significant impact on either DOM removal or THM and HAA_9 formation.

These results also show the role of DOM characteristics in the DOM adsorption by GAC and DBP control. The characterization results of the Myrtle Beach and Spartanburg waters show relatively similar $SUVA_{254}$ values, MW_w and % hydrophobicity. Despite these similarities, it is evident that they exhibited significantly different adsorption patterns by the same steam-treated GACs. The results indicate that the GAC performance in DBP formation control is highly dependent on the nature of DOM.

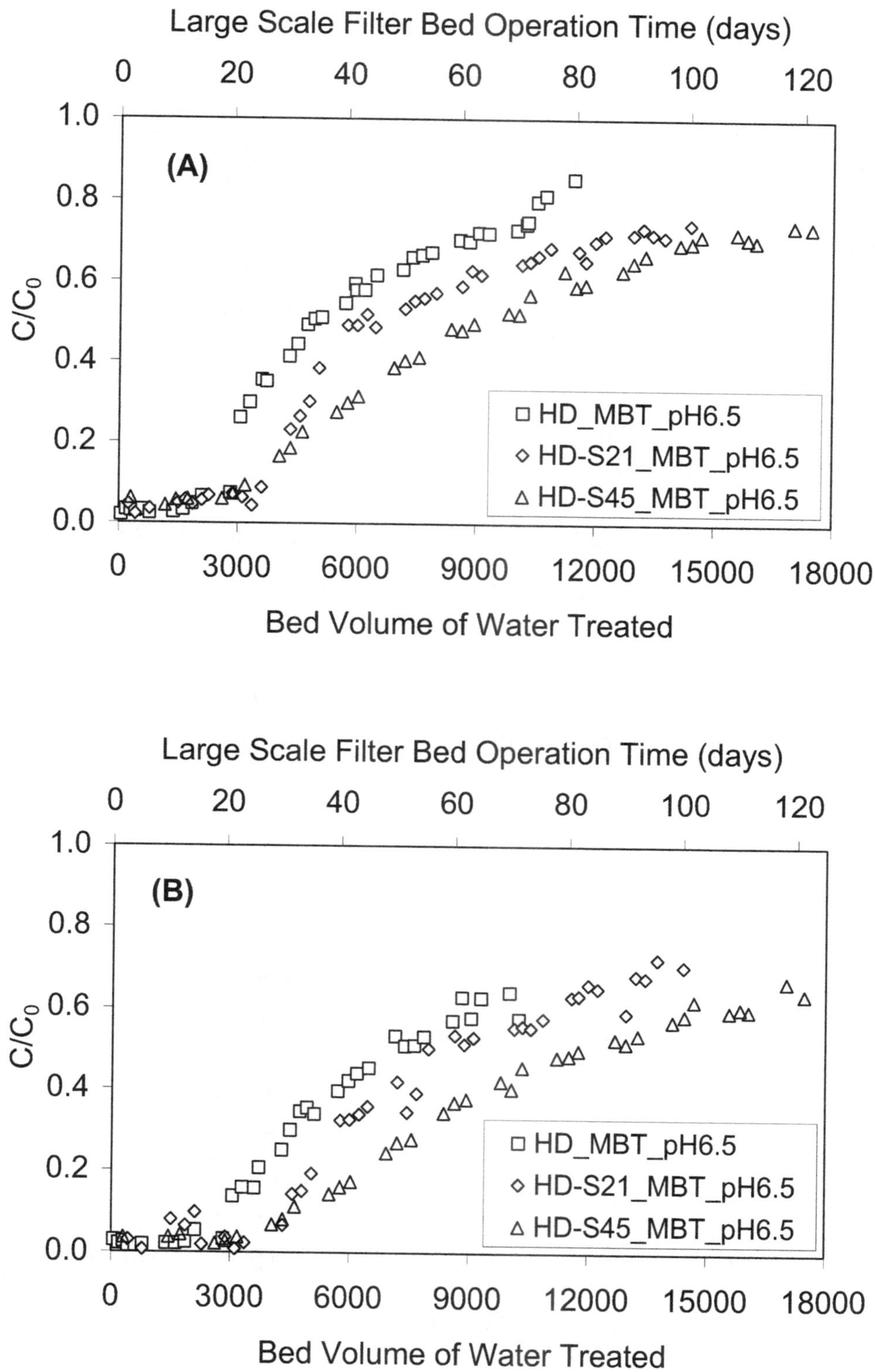

Figure 3.21 Myrtle Beach treated water DOM breakthrough profiles, as (A) TOC and (B) UV_{254}, for the virgin and steam-treated carbons

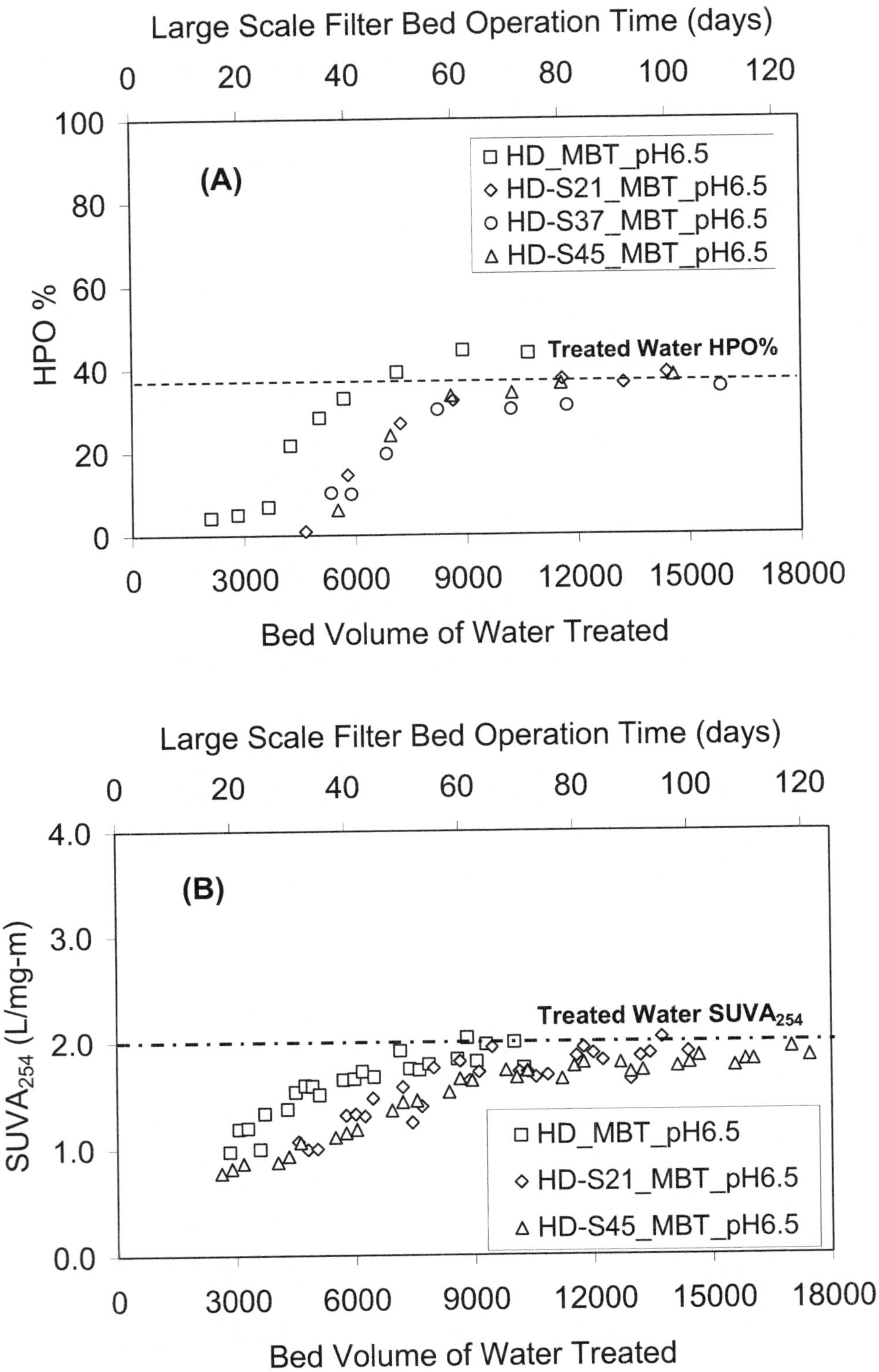

Figure 3.22 Myrtle Beach treated water DOM (A) %HPO and (B) $SUVA_{254}$ breakthrough profiles for the virgin and steam-treated GACs

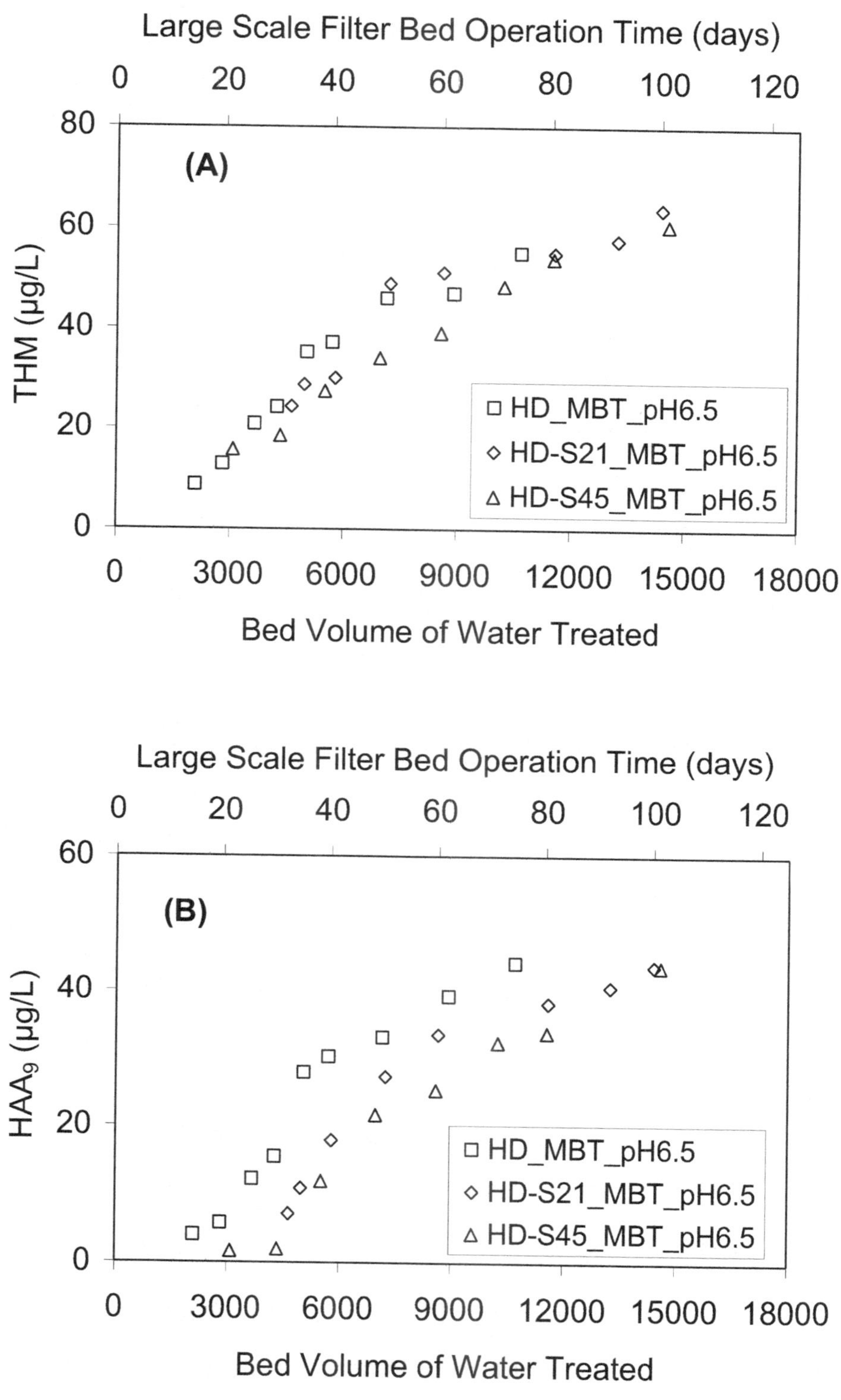

Figure 3.23 Myrtle Beach treated water DOM: (A) THM and (B) HAA_9 formation breakthrough profiles for virgin and steam-treated carbons

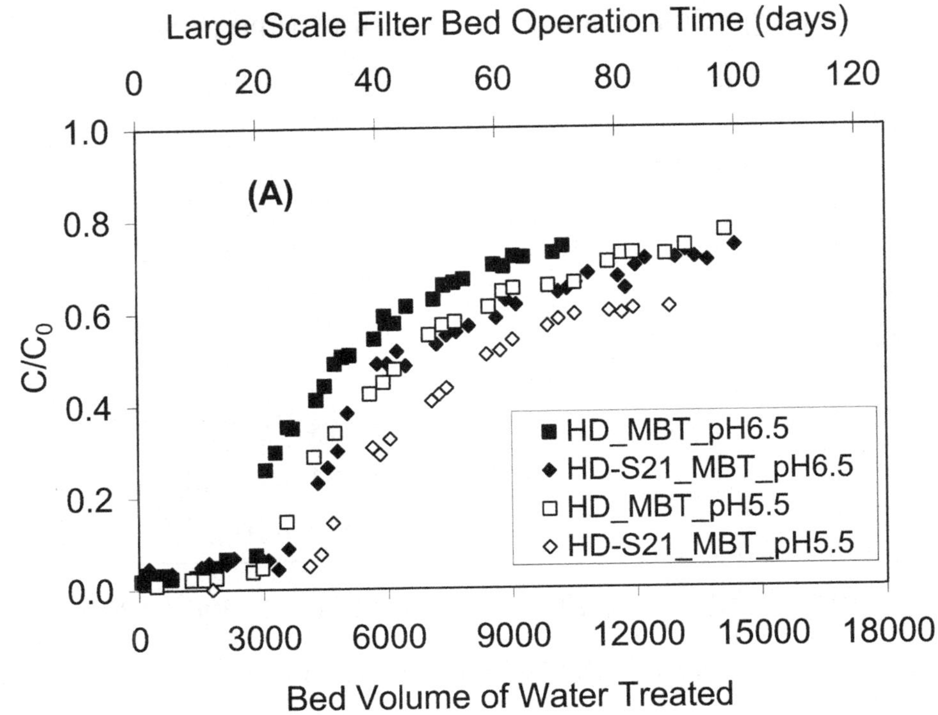

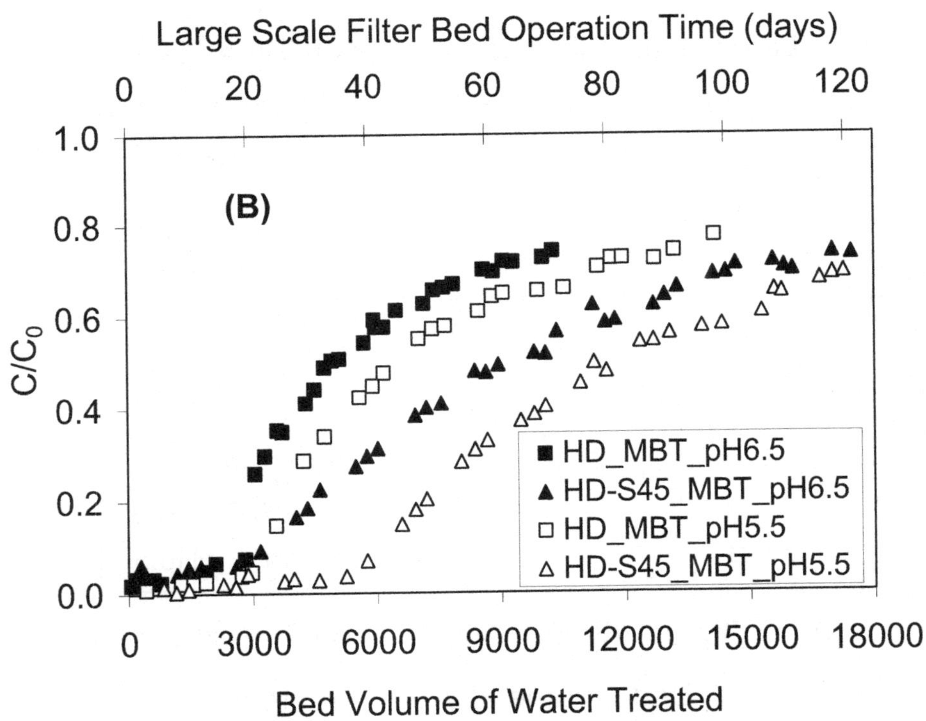

Figure 3.24 Myrtle Beach treated water DOM (TOC) breakthrough profiles for (A) HD-S21 and (B) HD-S45 carbons at pH 6.5 and 5.5

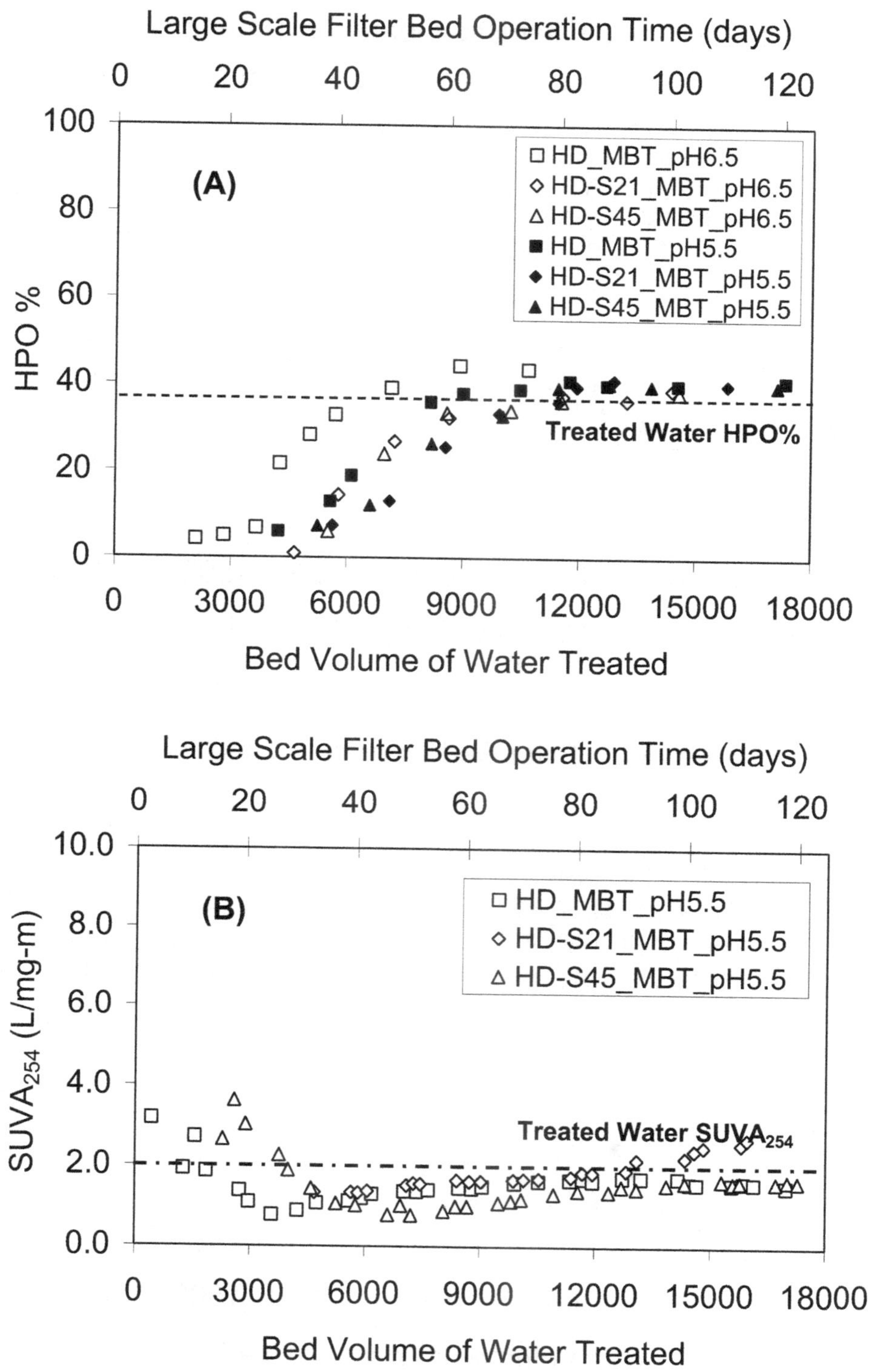

Figure 3.25 pH effect on Myrtle Beach treated water DOM (A) %HPO and (B) SUVA254 breakthrough profiles for the virgin and steam-treated GACs

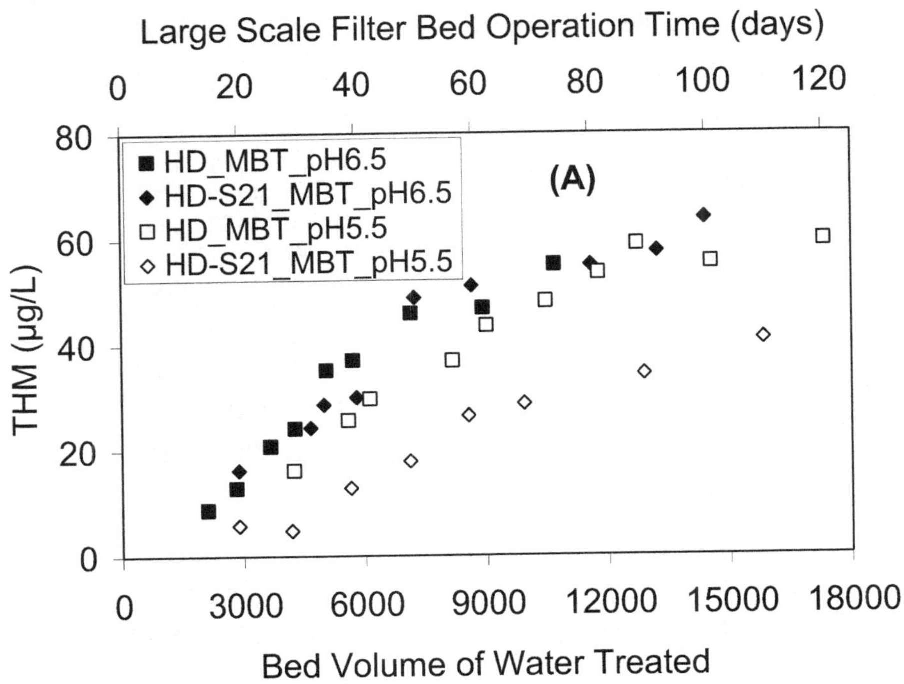

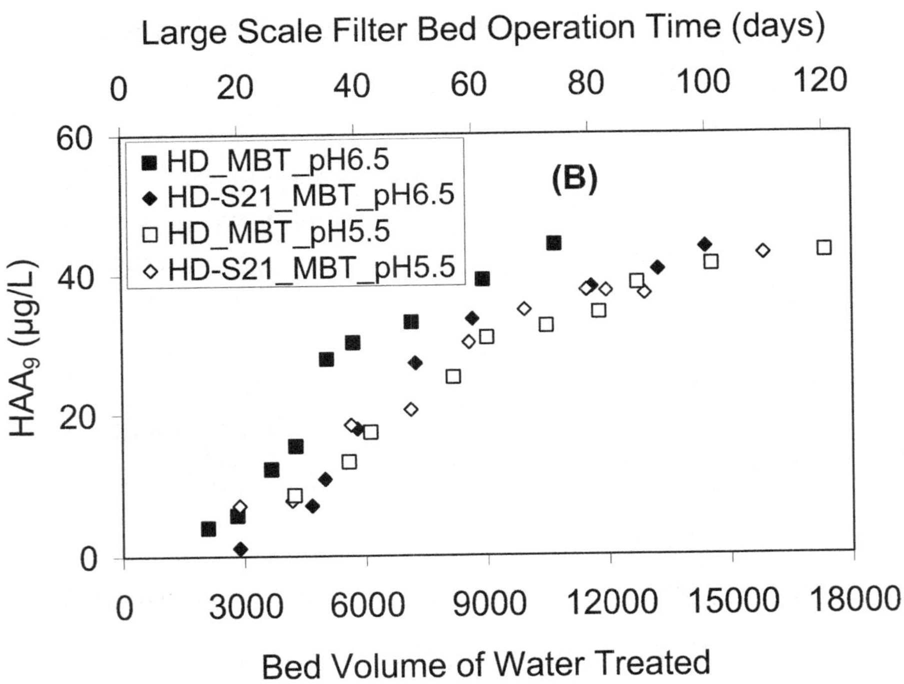

Figure 3.26 Myrtle Beach treated water DOM (A) THM and (B) HAA_9 formation for breakthrough samples of HD-S21 carbon at pH 6.5 and 5.5

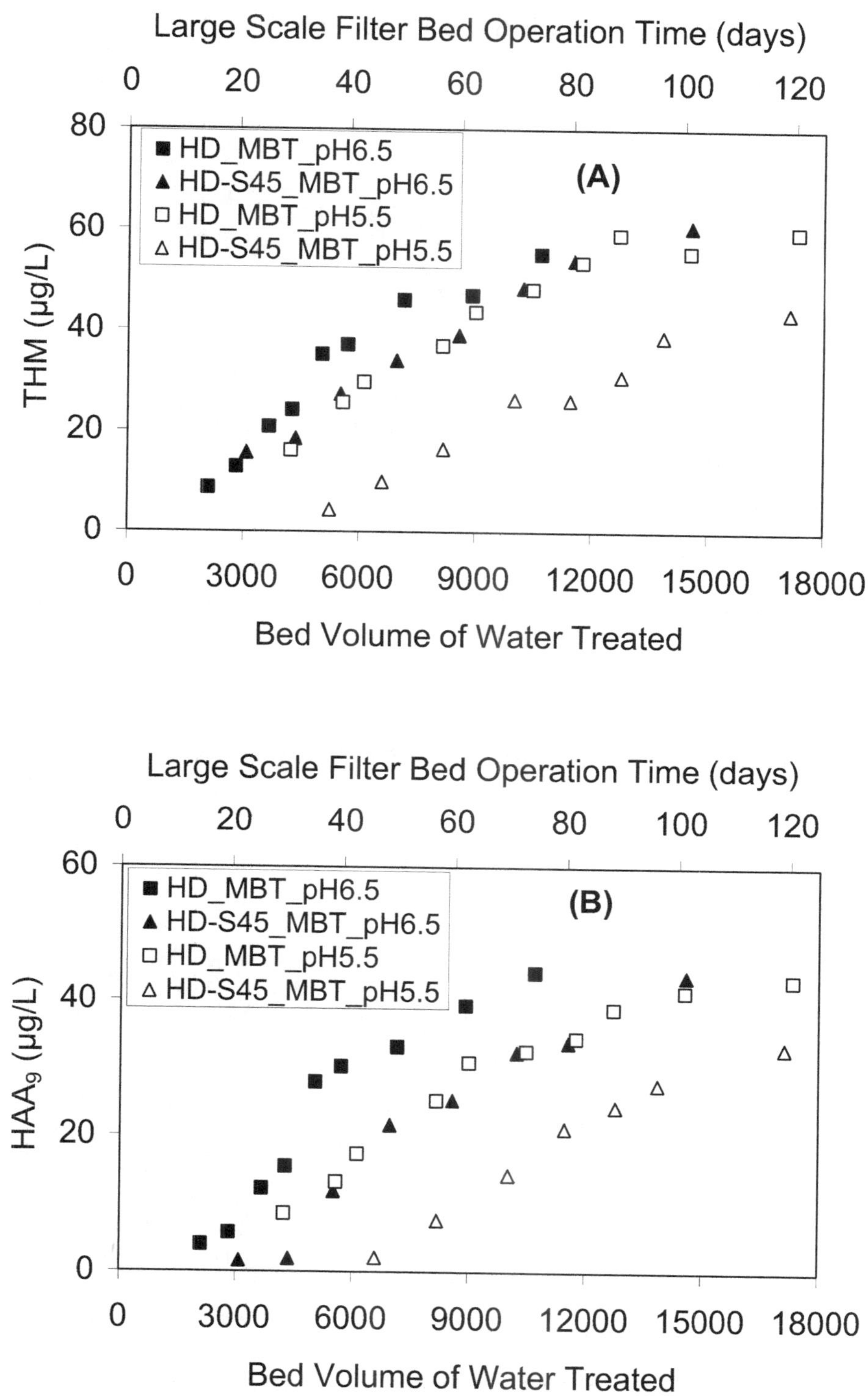

Figure 3.27 Myrtle Beach treated water DOM (A) THM and (B) HAA_9 formation for breakthrough samples of HD-S45 carbon at pH 6.5 and 5.5

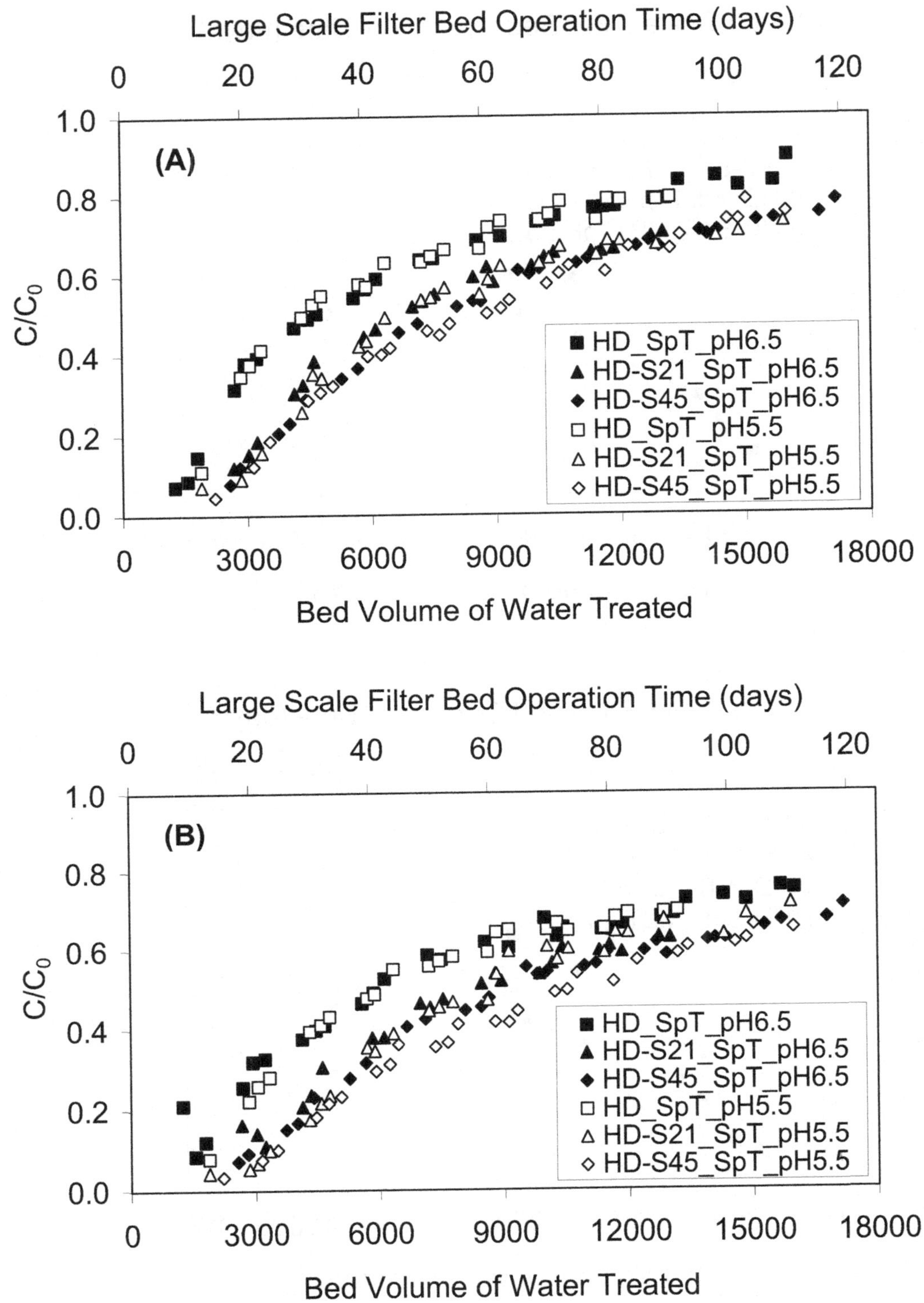

Figure 3.28 Spartanburg treated water DOM breakthrough profiles, as (A) TOC and (B) UV_{254}, for the steam-treated carbons at pH 6.5 and 5.5

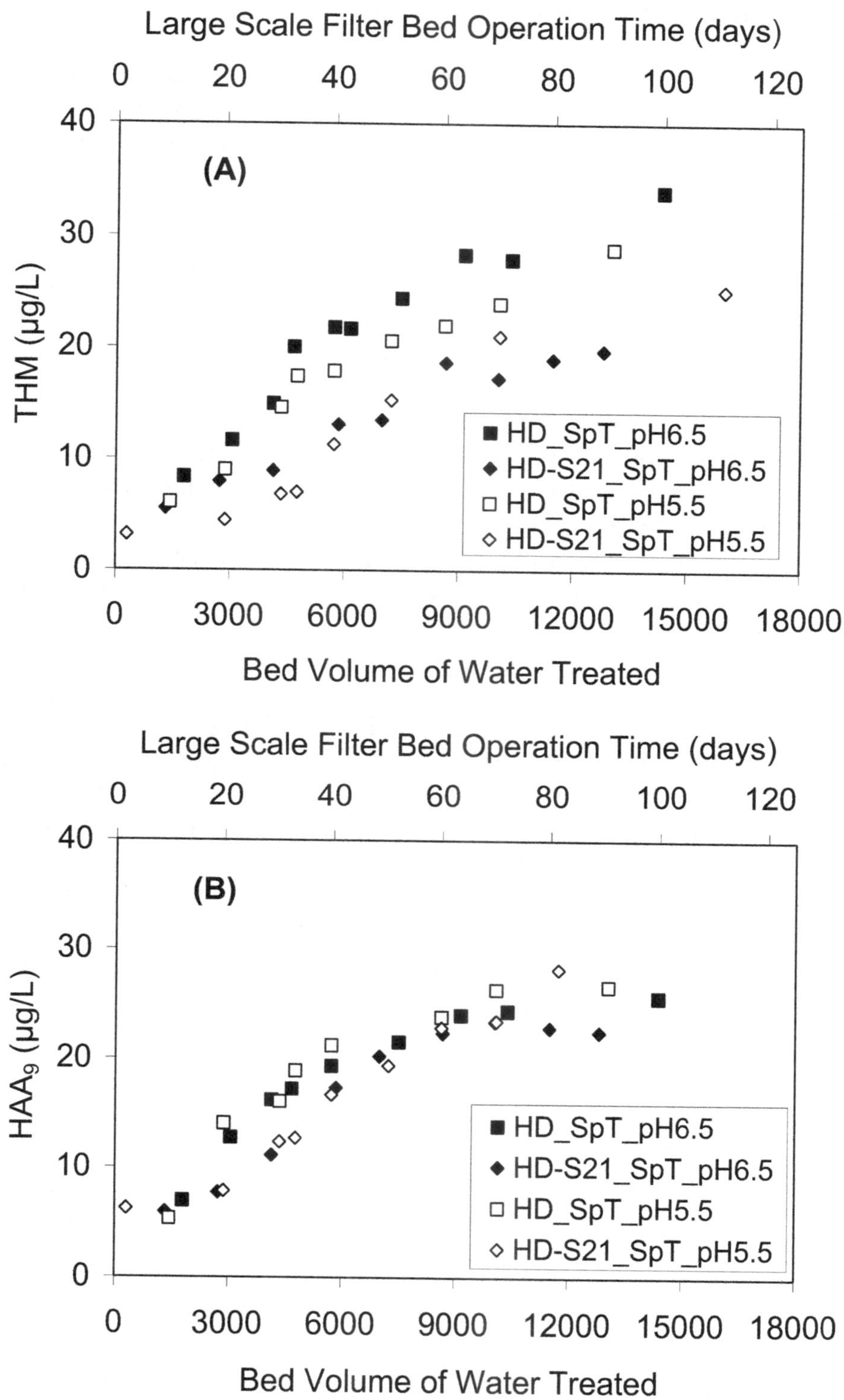

Figure 3.29 Spartanburg treated water DOM (A) THM and (B) HAA_9 formation for breakthrough samples of HD-S21 carbon at pH 6.5 and 5.5

Comparison of DOM uptake during isotherm and RSSCT

Calculations were performed to compare the DOM uptake in batch and RSSCT experiments for both the Myrtle Beach and Spartanburg waters. The DOM uptake of each carbon was calculated using the portion of the TOC breakthrough curves up to 70% (i.e., C/C_0=0.7), the flow rate, and the carbon mass in each column. The results are listed in Table 3.7. The DOM uptake calculated from the column experiments, although it was only up to 70% of the breakthrough, was significantly higher than the uptake obtained from the isotherms for the treated waters; however the difference was negligible for the raw waters. This difference is probably due to the heterogeneous nature of DOM and the different adsorption patterns between column and isotherm experiments. During a column experiment, the influent DOM concentration and composition were kept almost constant. The availability of strongly adsorbing DOM components was not a limiting factor, and activated carbon was always in contact with a 3 mg TOC/L solution. In contrast, during an isotherm experiment, there was a constant amount of DOM in an isotherm bottle. Initially strongly adsorbing components adsorbed, which was followed by adsorption of weakly adsorbing DOM components. Therefore, the composition of adsorbed DOM is different at the same equilibrium concentration between a column and isotherm experiment. The higher DOM uptake observed for the treated water in RSSCT experiments as compared to batch isotherms is probably because the capacity of activated carbon was used much more effectively by the strongly adsorbing components when the influent concentration was continuously kept constant at 3 mg/L. On the other hand, the similar DOM uptake for the raw water between the RSSCT and isotherm experiments indicates that the amount of strongly adsorbing DOM components (probably low molecular weight DOM) in the raw water was low. The overall DOM uptake was controlled mainly by the extent of adsorption of large DOM components. Therefore, no significant difference was observed between the uptake during both isotherm and column experiments of the raw waters.

These findings demonstrate that although, from a practical point of view, batch experiments are simple to screen various activated carbons for DOM removal, a dynamic RSSCT, despite its longer and more costly experimental nature, should be performed to obtain more representative information about the effectiveness of activated carbons for practical fixed-bed filter applications.

Table 3.7
Bed volume of water treated and DOM uptake by the virgin and steam-treated carbons for raw and treated Spartanburg water until 70% of TOC breakthrough, and the DOM uptake at C_e=3 mg/L from the isotherm experiments

Water	pH	Carbon	Up to 70% Breakthrough		Isotherm at Ce=3 mg/L
			Bed Volume of Water Treated	DOM uptake mg TOC/g GAC	DOM uptake mg TOC/g GAC
MBR	6.5	HD	3800	12.4	11.2
		HD-S21	5760	18.0	11.9
MBT	6.5	HD	8580	31.8	18.8
		HD-S21	12000	44.9	23.6
		HD-S45	14430	54.2	26.9
	5.5	HD	11340	43.5	23.8
		HD-S21	15960	60.4	29.3
		HD-S45	18120	74.1	36.0
SpR	6.5	HD	4590	15.4	16.3
SpT	6.5	HD	9120	30.6	19.8
		HD-S21	13020	44.6	24.6
		HD-S45	14100	45.4	26.4
	5.5	HD	8820	32.0	29.4
		HD-S21	14850	51.6	37.4
		HD-S45	13440	47.5	41.6

Regeneration Results

In this study, the effectiveness of activated carbon regeneration was investigated for the activated carbons collected from the Myrtle Beach treated water isotherm experiments at two initial concentrations, 20 mg TOC/L and 10 mg TOC/L. These carbons were regenerated two times after the first use. Heat regenerated carbon under H_2 was only performed on carbon loaded with 10 mg TOC/L solutions to compare with its counterpart under N_2. The DOM uptakes by the fresh and regenerated carbons are listed in Table 3.8.

High temperature regeneration under H_2 or N_2 atmosphere fully recovered GAC adsorption capacities after two regeneration cycles. There was no systematic pattern of decrease in the DOM uptake of steam-treated carbons with increasing regeneration cycles (Table 3.8). The gas type used for regeneration did not make a difference. However, it should be noted that the adsorption experiments were performed in relatively short period time after each regeneration. Aging (i.e., oxygen adsorption and development of surface acidity) of N_2 regenerated GACs is expected to be faster as compared to H_2 regenerated GACs. The weight loss during the heat treatment regeneration was ~6% for each carbon. The carbon characterization results after the last regeneration cycle did not show major changes in physical characteristics of the carbons (Table 3.9), which was consistent with the similar DOM uptake shown by the GACs after two regeneration cycles.

In contrast to the heat regenerated carbons, the NaOH regenerated carbons showed a continuous trend of decreasing DOM uptake with increasing regeneration cycle. After the first regeneration cycle, the losses of capacity for HD-S21-B and HD-S45-B were 17% and 18% at initial concentration of 10 mg TOC/L, and 26% and 20% at 20 mg TOC/L, respectively. After the second regeneration cycle, the total losses increased to 31% and 37% at 10 mg TOC/L, and 36% and 26% at 20 mg TOC/L, respectively. Characterization results did not show a major difference between high temperature and NaOH regenerated GAC, especially in pores >1 nm (Table 3.9). The lower capacities by the NaOH regenerated GACs as compared to the high temperature regenerated GACs were attributed to the difference in carbon surface chemistry. Under high temperature conditions, the adsorbed DOM molecules are expected to decompose on the carbon surface. However, after the NaOH regeneration, the remaining DOM on the carbon surface is expected to increase the negative charge on the surface. Therefore, at the pH condition of isotherms, these remaining DOM on the NaOH regenerated GACs may hinder adsorption of DOM. Since there was limited amount of carbon mass for the regeneration experiments, it was not possible to characterize the change in the surface chemistry of the regenerated carbons.

Table 3.8
Myrtle Beach treated water DOM uptakes of fresh and regenerated steam-treated activated carbons

Carbon*	1st adsorption isotherm			2nd adsorption isotherm			3rd adsorption isotherm		
	Initial mg/L	After mg/L	Uptake mg TOC/g	Initial mg/L	After mg/L	Uptake mg TOC/g	Initial mg/L	After mg/L	Uptake mg TOC/g
HD-S21-H	10.1	2.2	15.7	9.6	1.0	17.1	9.6	1.5	16.1
HD-S21-N	10.1	2.2	15.7	9.6	1.0	17.2	9.6	2.0	15.2
HD-S21-B	10.1	2.2	15.7	9.6	3.1	13.1	9.6	4.2	10.9
HD-S45-N	10.1	2.1	16.0	9.6	1.2	16.7	9.6	2.1	15.1
HD-S45-B	10.1	2.1	16.0	9.6	3.0	13.1	9.6	3.7	11.8
HD-S21-N	19.9	6.3	27.1	19.3	6.3	26.0	19.3	7.0	24.6
HD-S21-B	19.9	6.3	27.1	19.3	9.3	20.0	19.3	10.4	17.8
HD-S45-N	19.9	6.0	27.7	19.3	5.6	27.3	19.3	5.9	26.8
HD-S45-B	19.9	6.0	27.7	19.3	8.3	22.1	19.3	9.1	20.5

* HD-SXX-H: GAC regenerated under H_2 atmosphere, HD-SXX-N: GAC regenerated under N_2 atmosphere, and HD-SXX-B: GAC regenerated with NaOH

Table 3.9
Characterization of fresh and regenerated steam-treated activated carbons after 2nd regeneration cycle

Sample Name	S_{BET} (m^2/g)	V_t (cm^3/g)	V_{micro} (cm^3/g)	$V_{meso+macro}$ (cm^3/g)	DFT Surface Area Distribution (m^2/g) <100nm	<1 nm	>1 nm	1-2 nm	2-3 nm	>3 nm
HD-S21	853	0.79	0.325	0.465	479	282	197	81	28	88
HD-S21-M10-H2	794	0.712	0.288	0.424	579	373	206	97	30	84
HD-S21-M10-N2	782	0.746	0.284	0.462	575	371	204	87	32	88
HD-S21-M10-B2	841	0.773	0.31	0.463	618	421	197	114	23	60
HD-S21-M20-N2	830	0.803	0.302	0.501	573	377	196	97	27	71
HD-S21-M20-B2	741	0.675	0.269	0.406	536	332	204	93	18	83
HD-S45	1078	0.947	0.399	0.548	703	423	280	151	48	81
HD-S45-M10-N2	845	0.686	0.308	0.378	618	407	201	102	37	72
HD-S45-M10-B2	974	0.8	0.351	0.449	702	435	267	138	43	86
HD-S45-M20-N2	893	0.742	0.319	0.423	631	385	246	119	42	85
HD-S45-M20-B2	943	0.804	0.342	0.462	646	411	235	131	36	68

SUMMARY OF IMPORTANT FINDINGS

The following important findings were obtained in this phase of the study:

- The results obtained in this study using both raw and treated waters at the typical pH range of water treatment (i.e., pH 5.5 to 8) provided direct experimental evidence that the great majority of DOM components in natural waters cannot access pores smaller than 1 nm. Therefore, when a GAC to be selected for DOM removal, GACs with large surface areas and volumes in pores >1 nm should be preferred. The BET surface area and the total pore volume are not the important parameters in selecting GAC for DOM removal because they do not provide information about the *accessible* pore region. GAC surface chemistry also plays a role in DOM adsorption. It was shown that the presence of high surface acidity (i.e., acidic pH_{PZC} values) significantly decreases DOM adsorption. However, the effect of pore size distribution becomes more important than the pH_{PZC} at the neutral or basic pH_{PZC} values.
- GACs with wider pore size distribution and higher pH_{PZC} can be produced by extending steam treatment periods during carbon preparation. The steam-treated GACs showed higher DOM uptakes compared with their precursor. However, the increase in DOM uptake was not always proportional to the extent of steam treatment. The increase uptake of DOM is due to removal of DOM components with more hydrophobic character and having higher $SUVA_{254}$ values.
- The physicochemical characteristics of DOM play important roles in the DOM uptake. Although the higher accessible pore volume and higher pH_{PZC} as a result of steam treatment resulted in higher uptake of the Myrtle Beach treated water DOM by HD-S45 than HD-S21, the difference between the Spartanburg treated water DOM adsorption by HD-S45 and HD-S21 was negligible. For both waters, HD-S21 exhibited significantly higher DOM adsorption than the virgin HD. Opening carbon pores by steam treatment enhanced the DOM uptake, however extended steam treatment and additional pore enlargement did not result in additional DOM adsorption for the Spartanburg treated water DOM. Myrtle Beach and Spartanburg treated DOM had similar $SUVA_{254}$ and MW_w values determined through SEC. The increase in the DOM uptake was not found to correlate with these simple DOM characterization parameters. It appears that there were differences (e.g., the *size* and structure) between these two DOM mixtures that were not fully resolved by $SUVA_{254}$ and MW_w measurements, causing differences in their adsorption by porous activated carbons.
- The RSSCT results for Myrtle Beach and Spartanburg water DOM at pH 6.5 and 5.5 indicated four conclusions. (1) Steam-treated carbons had a longer operation time than their precursor to reach a certain degree of breakthrough, (2) The pH effect on DOM removal depends on the source water characteristics. For the high $SUVA_{254}$, more hydrophobic Myrtle Beach DOM, decrease in pH increased the DOM uptake by GACs significantly. While for the low $SUVA_{254}$, more hydrophilic Spartanburg DOM, the pH effect was minimal. This was attributed to the less impact of pH on the change of molecular size or conformation of hydrophilic DOM in Spartanburg water, (3) Generally, the steam-treated GACs can increase the removal of THM and/or HAA_9 precursors. However, the extent of removal was water specific and pH dependent. The results obtained indicate that GACs with large mesopore volume and high pH_{PZC} will be more effective for DBP control in high SUVA waters. In this study, for the Myrtle Beach treated water, after 18,000 bed volumes and approximately 4+ months of operation, the steam-treated HD-S45 at pH 5.5 had effluent THM and HAA_9

formations well below the regulatory limits, 80 and 60 ppb, respectively, and (4) UV absorbing DOM molecules mainly contributed to the DBP formation in water treatment.

- The comparison of DOM uptake in isotherm tests and RSSCTs showed that although, from a practical point of view, an isotherm experiment is a simple technique that can be used to screen various activated carbons for DOM removal, a dynamic RSSCT, despite its longer and more costly experimental nature, should be performed to obtain representative information about the effectiveness of activated carbons for practical applications. From a DBP control perspective, RSSCT should also be combined with DBP formation testing to determine the feasibility of GAC adsorption for DBP control. Monitoring TOC removal alone is not conclusive about the effectiveness of GAC adsorption for DBP formation control.
- Regeneration experiments showed that high temperature regeneration under H_2 or N_2 atmospheres effectively restored the DOM adsorption capacity of steam-treated GACs for two cycles of regeneration. In contrast, the NaOH regenerated carbons showed a continuous trend of decreasing DOM uptake with increasing regeneration cycle. It appears that in practice high temperature steam treatment is a more feasible choice for carbon regeneration

CHAPTER 4
ADSORPTION OF DOM AND CONTROL OF DBP FORMATION BY IRON-IMPREGNATED GACS

INTRODUCTION

The results of batch adsorption tests in the previous phases of this project showed that iron-impregnated activated carbons demonstrated enhanced DOM adsorption (Chapters 1 and 2). This was attributed to the formation of complexes between the iron species on the carbon surface and some DOM components. The main objectives in this phase were to further investigate the effectiveness of iron-impregnated GACs for DOM adsorption and especially to examine the control of THM and HAA formation using iron-impregnated GACs in practical applications. Various metal loading and activation techniques, which are widely used in catalyst preparation, were tested to deposit iron species on activated carbon surface. Furthermore, emphasis was placed on RSSCTs to examine the effectiveness of iron-impregnated GACs in practical applications for the control of DBP formation.

MATERIALS AND METHODS

Adsorbents

The HD-S21 carbon prepared and described in Chapter 3 was used as the precursor in the preparation of iron-impregnated GACs. Various iron loading methods, which are commonly used in the catalyst preparation, are explored in this study. Following section will provide a brief description of these methods.

Excess solution impregnation followed by heat treatment

The excess solution impregnation method was used to load ionic iron species on the surfaces of HD-S21. In this procedure, 208 g of $FeCl_3 \cdot 6H_2O$ was initially dissolved in 400 mL of DDW. After purging with nitrogen for about 15 minutes to remove the dissolved oxygen, approximately 175 g of HD-S21 were mixed with this solution for several hours while stirring and purging with N_2. The water was then removed from the mixture by evaporation. The carbon was subsequently heated at 800°C for 2 hours under nitrogen flow. The carbon obtained with this process was labeled as HD-S21-FeS.

Ionic iron loading followed by sodium hydroxide precipitation

A solution of $FeCl_3$ (dissolving 155 g of $FeCl_3 \cdot 6H_2O$ in 1 L DDW) was used as the iron precursor. In this method, 160 g of HD-S21 were mixed with the $FeCl_3$ precursor solution, and the pH of the slurry was adjusted to 7, then to 9 with NaOH solution while stirring. After stirring for 30 minutes, the water was evaporated by heating and the carbon was dried in an oven at 80~90°C. The produced carbon was labeled as HD-S21-FeOH.

Organic iron loading followed by heat treatment

An organic iron compound was used as the iron source. The precursor solution was prepared by dissolving 193 g of iron (II) oxalate dihydrate ($FeC_2O_4 \cdot 2H_2O$) in about 300 mL concentrated HCl (36.5 wt %). During the impregnation step, 150 g of HD-S21 were mixed with the precursor solution and soaked for several hours. After the solvent was evaporated by heating, the carbon was heat treated at 800°C for 2 hours under nitrogen flow. The sample obtained after this procedure was labeled as HD-S21-FeOX.

Ionic iron loading followed by $NaBH_4$ reduction

The iron precursor was obtained by dissolving 183 g of $FeCl_3 \cdot 6H_2O$ in 1 L DDW. Iron loading was performed through excess solution impregnation followed by solvent evaporation and then $NaBH_4$ reduction. HD-S21 was initially soaked in the iron precursor solution for 20 minutes. Then, 500 mL of $NaBH_4$ solution (containing 89 g of $NaBH_4$) were added drop wise to the carbon slurry while stirring. After stirring for 2 hours, the carbon was filtered and dried in an oven at 90°C overnight. The sample obtained with this process was labeled as HD-S21-Fe^0.

Nitric acid oxidation followed by ionic iron loading and heat treatment

In this preparation, to improve iron dispersion as well as increase the iron content on the carbon surface, more ion exchange sites (acidic groups) were initially introduced onto the surface of HD-S21 by boiling the GAC in concentrated nitric acid for 1 hour. Then iron was loaded on the surface of the oxidized carbon by contacting 15 g of HD-S21 with $FeCl_3$ solution consisting of 12 or 24 g $FeCl_3 \cdot 6H_2O$ in 100mL DDW. The carbon was separated from solution by filtration and dried in an oven at 80~90°C and heat treated at 800°C for 2 hours under nitrogen flow to decompose $FeCl_3$ crystals and form iron oxide species. The carbons produced with this method were labeled as HD-S21-FeI1 (using 12 g of $FeCl_3$) and HD-S21-FeI2 (using 24 g of $FeCl_3$), respectively.

Washing and sieving

After each modification procedure, carbons were thoroughly washed with DDW until there was no change in the pH of the filtrate, and no iron could be detected in the filtrate for iron-impregnated GACs. Subsequently, the GACs were dried in an oven at 80~90°C, and the samples were carefully crushed with mortar and pestle and sieved. The particles between mesh sizes of #100 and #140 were used in all experiments.

Characterization of adsorbents

The virgin and modified GACs were characterized using the methods described in Chapter 2.

DOM Samples

The same DOM batches collected for the steam-treated GACs in Chapter 3 were used in the experiments. This also allowed a side-by-side comparison of the performances of the steam-treated and iron-impregnated GACs.

Experimental and Analytical Methods

The details about the methods for adsorption isotherms, RSSCTs, batch mode DOM fractionation, DBP formation tests, and various analytical techniques used in this phase of the study were described previously in Chapters 2 and 3.

RESULTS AND DISCUSSION

Characteristics of Adsorbents

The characterization results of the iron-impregnated GACs and their precursor are provided in Table 4.1. The characterization results indicated that various amounts of iron species (1.23~8.73%) were loaded on the carbon surface through iron impregnation. For the iron-impregnated GACs, the change in pore volume and surface area, expressed on a mass basis, compared to the virgin carbon can result from a combination of various factors: (1) the weight increase as a result of iron loading; (2) the deposition of iron in carbon pores and/or at some of the pore entrances which are expected to change both the pore volume and size distribution; and (3) the modification procedures that include the final high temperature treatment step may create new pores and/or widen the existing pores due to the volatilization of carbon materials catalyzed by iron species. It was assumed that the contribution of the iron species to the total carbon surface area was negligible

The impact of iron impregnation, except for excess solution impregnation followed by heat treatment (FeS), was, in general, to decrease the surface area, surface area in pores >1nm, and total pore volume of the activated carbons. It is clear that iron deposition resulted in a negative impact on the physical characteristics of activated carbons. However, the degree of impact was not proportional to the degree of iron loading. The pH_{PZC} values were lower as compared to HD-S21, except that HD-S21-FeOX and HD-S21-Fe^0 had similar values.

Table 4.1
Physicochemical characteristics of virgin and iron-impregnated GACs

Sample Name	pH_{pzc}	Fe content (%)	S_{BET} (m^2/g)	V_t (cm^3/g)	V_{micro} (cm^3/g)	$V_{meso+macro}$ (cm^3/g)	DFT Surface Area Distribution (m^2/g) <100nm	<1 nm	>1 nm	1-2 nm	2-3 nm	>3 nm
HD-S21	8.5	0.12	853	0.790	0.325	0.465	479	282	197	81	28	88
HD-S21-FeS	5.7	4.09	860	0.766	0.334	0.432	572	405	167	90	22	55
HD-S21-FeOH	7.5	1.23	832	0.752	0.321	0.431	599	422	177	72	27	78
HD-S21-FeOX	8.2	8.73	772	0.690	0.301	0.389	564	389	175	72	26	77
HD-S21-Fe^0	8.4	3.36	772	0.761	0.305	0.456	431	270	161	70	24	67
HD-S21-FeI1	5.4	3.16	836	0.783	0.31	0.473	455	272	183	91	26	66
HD-S21-FeI2	5.5	3.09	863	0.781	0.325	0.456	531	373	158	69	24	65

Adsorption Isotherms

Adsorption isotherms with the Myrtle Beach and Spartanburg DOM were obtained at pH 5.5 and 7.0. The results were expressed using both mass and surface area normalized units. Surface area based isotherms were used to determine the effect of iron on DOM adsorption. As discussed in Chapter 3, the surface area in pores larger than 1 nm was considered as accessible e area for the isotherm normalization.

Myrtle Beach treated water

Mass and surface-area based isotherms of the Myrtle Beach treated water DOM at pH 5.5 and 7 by the iron-impregnated carbons and their precursor, HD-S21, are shown in Figures 4.1 and 4.2, respectively. On a mass-basis, at an equilibrium concentration of 3 mg TOC/L or higher, the DOM uptake of the carbons were in the order of HD-S21-FeI1 ≈ HD-S21-FeI2 > HD-S21 ≈ HD-S21-FeOH > HD-S21-FeS ≈ HD-S21-FeOX > HD-S21-Fe^0 at both pH conditions. On a surface-area basis, the observed trends at both pH values appear to group into two categories from the highest to the lowest uptake. The first category consisted of HD-S21-FeI1 and HD-S21-FeI2; the second category included HD-S21, HD-S21-FeOH, HD-S21-FeS, HD-S21-FeOX and HD-S21-Fe^0. These trends did not correlate with the iron contents of the carbons. The total surface area in pores >1 nm were comparable for all iron-impregnated carbons (Table 4.1). The two best performing carbons HD-S21-FeI1 and HD-S21-FeI2 had about 3% iron content and the lowest pH_{PZC} values among all the iron-impregnated carbons. These observations indicate that the iron on the carbon surface interacts with some DOM components. However, the affinity of the impregnated activated carbons to DOM probably depends on the form of the iron and its dispersion on the carbon surface, and not necessarily the overall iron content. Although the characteristics of iron on the carbon surfaces were not determined in this study, which would be very difficult (may be impossible) to accomplish especially for iron dispersed in the carbon pores, it is reasonable to expect that different iron impregnation techniques employed in this study resulted in the formation of different iron species on the carbon surfaces. The most effective carbons, HD-S21-FeI1 and HD-S21-FeI2, were obtained with the ion-exchange method, where the precursor was initially oxidized. The initial oxidation step has been shown to introduce a significant amount of carboxylic groups, (i.e., the ion exchange sites) that can serve as anchor sites for iron and improve iron dispersion on the carbon surface (Radovic and Rodríguez-Reinoso, 1997; Rodríguez-Reinoso, 1998; Aksoylu et al., 2001). Decreasing the pH from 7 to 5.5 increased the DOM uptake by all carbons. The increase in the DOM uptake was 27%, 22%, 23%, 19%, 24%, 24% and 26% at C_e of 3 mg TOC /L for HD-S21, HD-S21-FeS, HD-S21-FeOH, HD-S21-FeOX, HD-S21-Fe^0, HD-S21-FeI1 and HD-S21-FeI2, respectively.

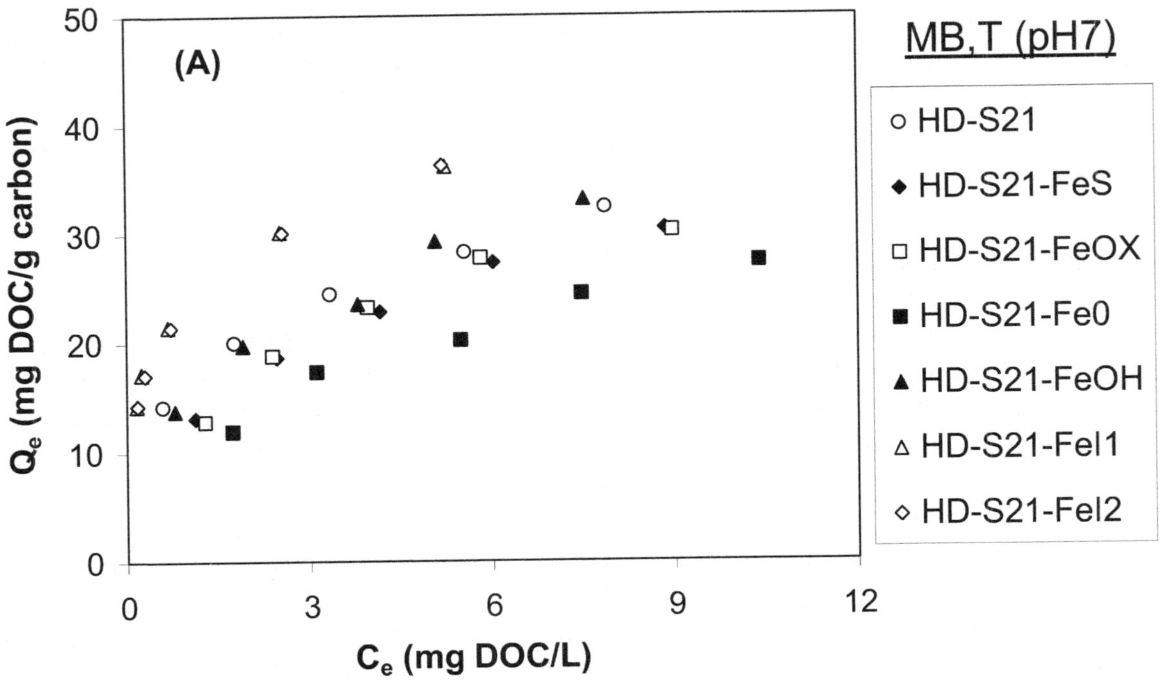

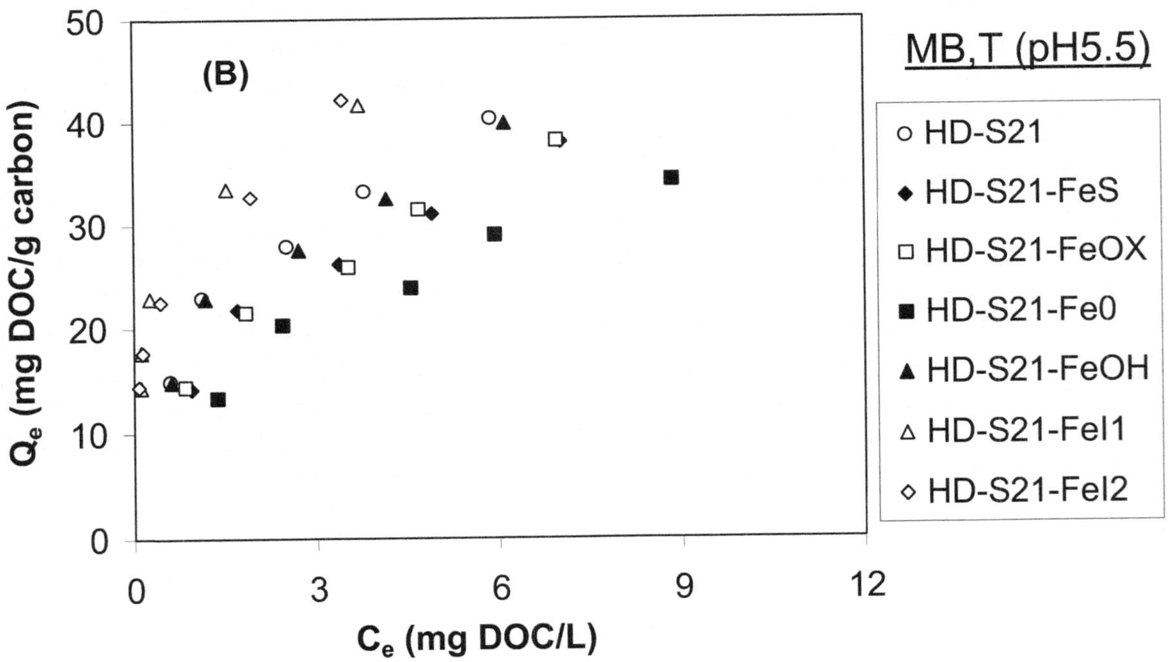

Figure 4.1 Mass basis isotherms of Myrtle Beach treated water DOM on HD, HD-S21 and iron impregnated carbons at: (A) pH of 7, (B) pH of 5.5. See text for explanation of legend codes.

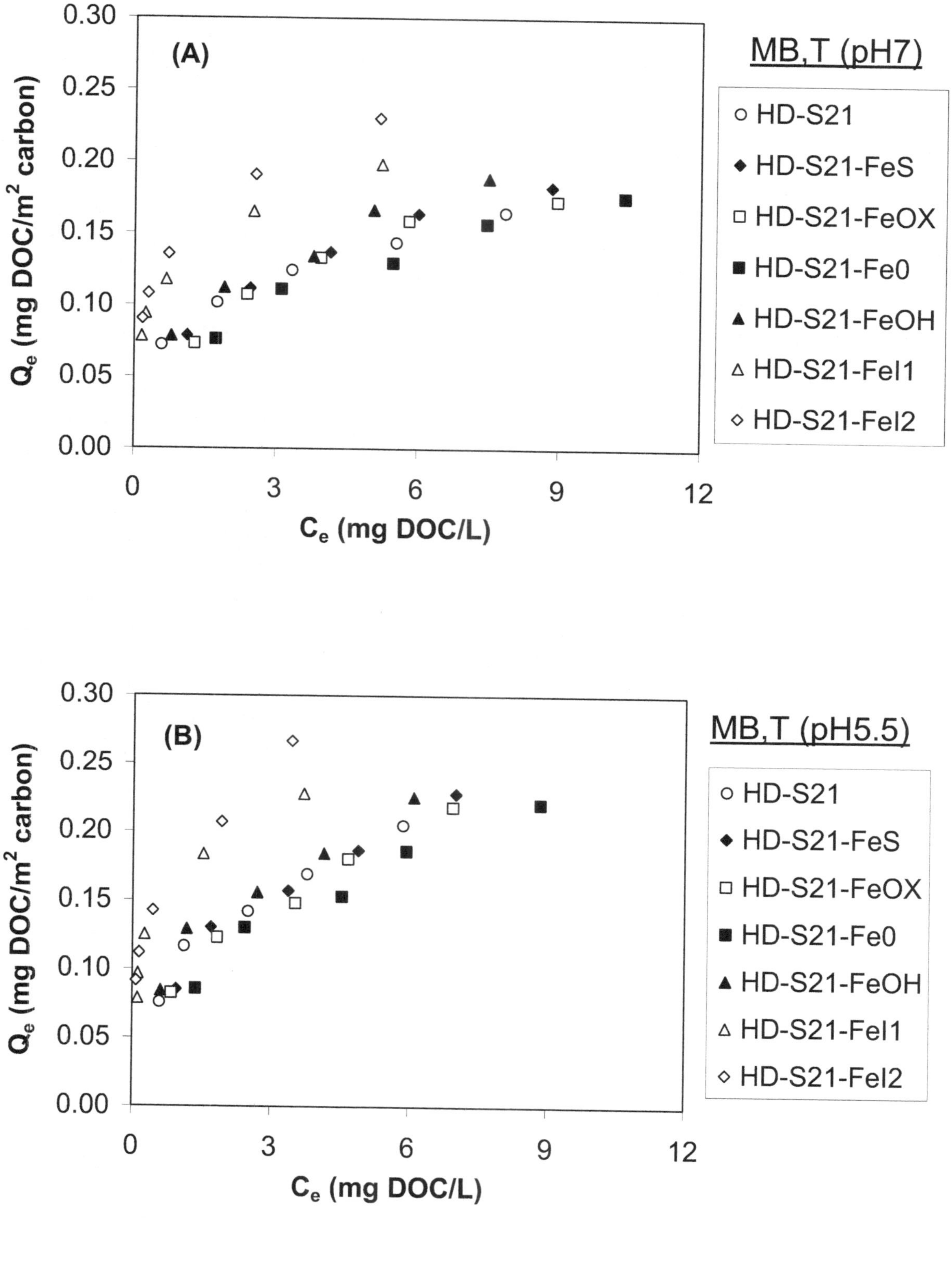

Figure 4.2 Surface-area basis isotherms of Myrtle Beach treated water DOM on HD, HD-S21 and iron impregnated carbons at: (A) pH of 7, (B) pH of 5.5

Myrtle Beach raw water

Mass and surface-area basis isotherms of Myrtle Beach raw water DOM at pH of 5.5 and 7 by iron-impregnated carbons and HD-S21 are shown in Figures 4.3 and 4.4, respectively. On a mass basis the iron-impregnated carbons showed different orders of DOM uptakes depending on pH. At pH 7, for C_e of 3 mg TOC/L, the mass-basis uptakes followed the order of HD-S21-FeI1 ≈ HD-S21-FeI2 > HD-S21 > HD-S21-FeOX ≈ HD-S21-FeOH ≈ HD-S21-FeS > HD-S21-Fe^0. On the other hand, at pH 5.5, the DOM uptake was in the order of HD-S21-FeI1 ≈ HD-S21-FeI2 > HD-S21 > HD-S21-FeOX ≈ HD-S21-FeOH > HD-S21-FeS > HD-S21-Fe^0. It was noteworthy that the best performing iron-impregnated GACs, HD-S21-FeI1 ≈ HD-S21-FeI2, showed much higher uptakes in treated than in raw water. This indicates that there were some components that were reactive in the Myrtle Beach DOM and the amounts of these components were higher in the treated than in the raw water given the fact that both raw and treated water were performed in the same concentration range.

On a surface-area basis, the order of DOM uptake at both pH values was HD-S21-FeI1 ≈ HD-S21-FeI2 > HD-S21 ≈ HD-S21-FeOH ≈ HD-S21-FeOX ≈ HD-S21-FeS > HD-S21-Fe^0. In both raw and treated waters the same iron-impregnated carbons, HD-S21-FeI1 and HD-S21-FeI2, showed the highest DOM uptake.

When solution pH decreased from 7 to 5.5, the order of relative increase in the DOM uptake was HD-S21, HD-S21-FeS, HD-S21-FeOH, HD-S21-FeOX, HD-S21-Fe^0, HD-S21-FeI1 and HD-S21-FeI2 was 93%, 65%, 93%, 91%, 117%, 70%, and 69%, respectively, at C_e of 3 mg TOC/L. This variable degree of pH impact on the different carbons suggests different (and more heterogeneous) DOM components interacted with the different forms of iron species on the carbon surface. At the same pH, the DOM uptake from the treated water was higher than the raw water for each carbon. The difference between the treated and raw water DOM uptakes on the same carbon was higher at pH of 7 than at pH of 5.5. This was attributed to a smaller difference between the molecular sizes of the raw and treated water DOM at pH 5.5 than 7 due to the coiling effect.

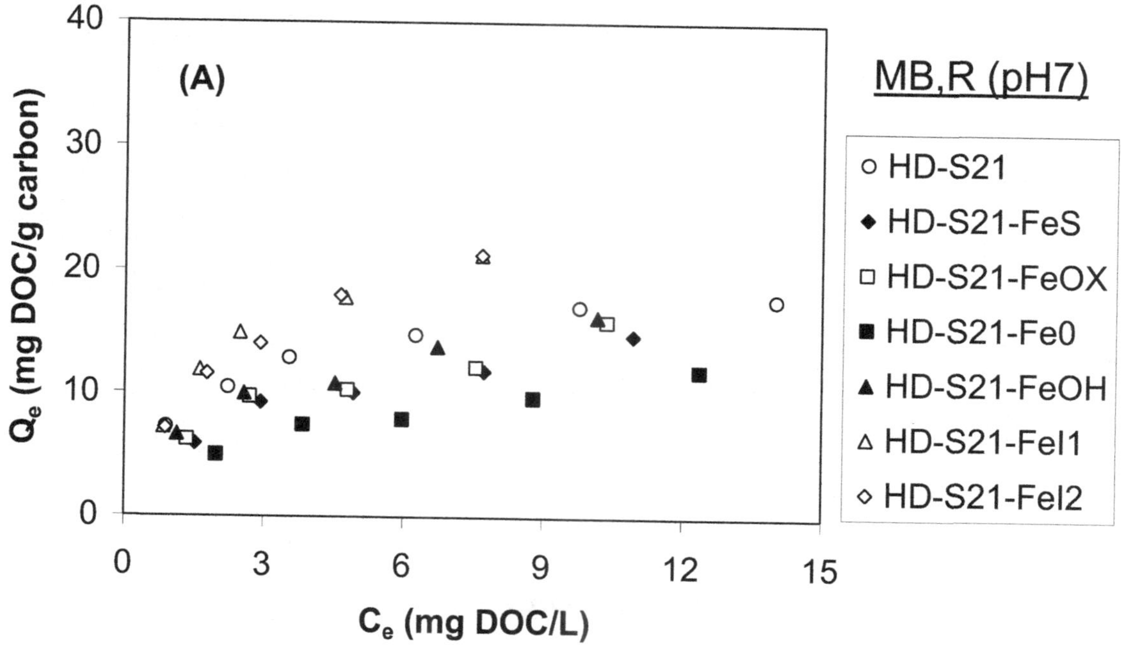

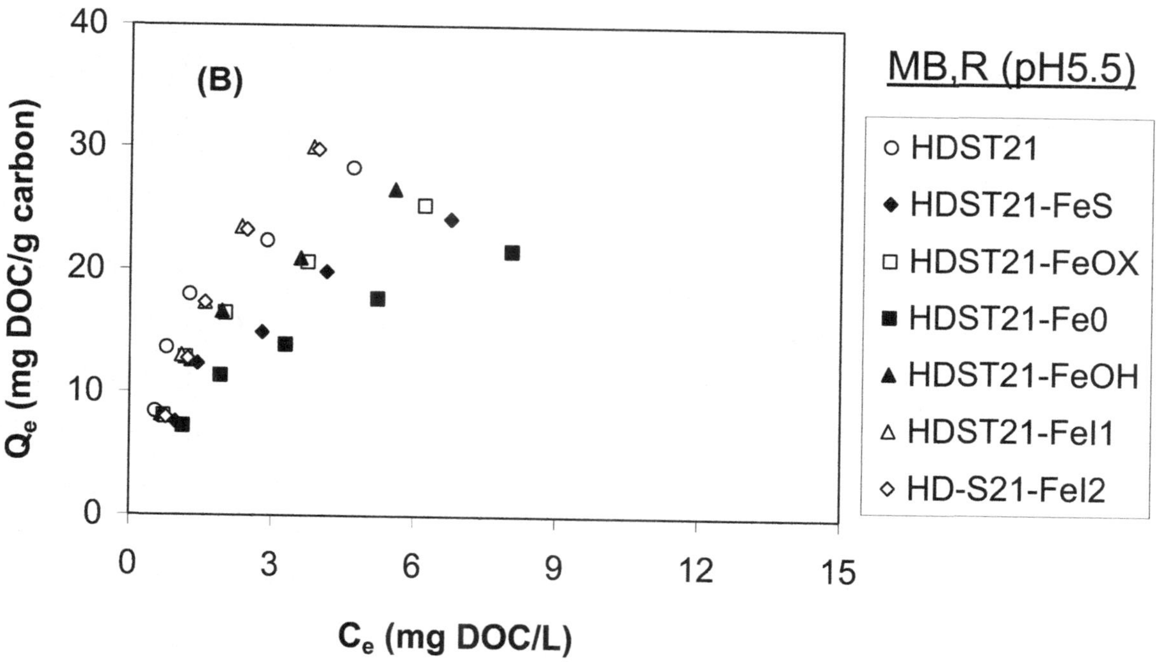

Figure 4.3 Mass basis Myrtle Beach raw water DOM adsorption isotherms of HD, HD-S21 and iron-impregnated carbons at: (A) pH of 7, (B) pH of 5.5

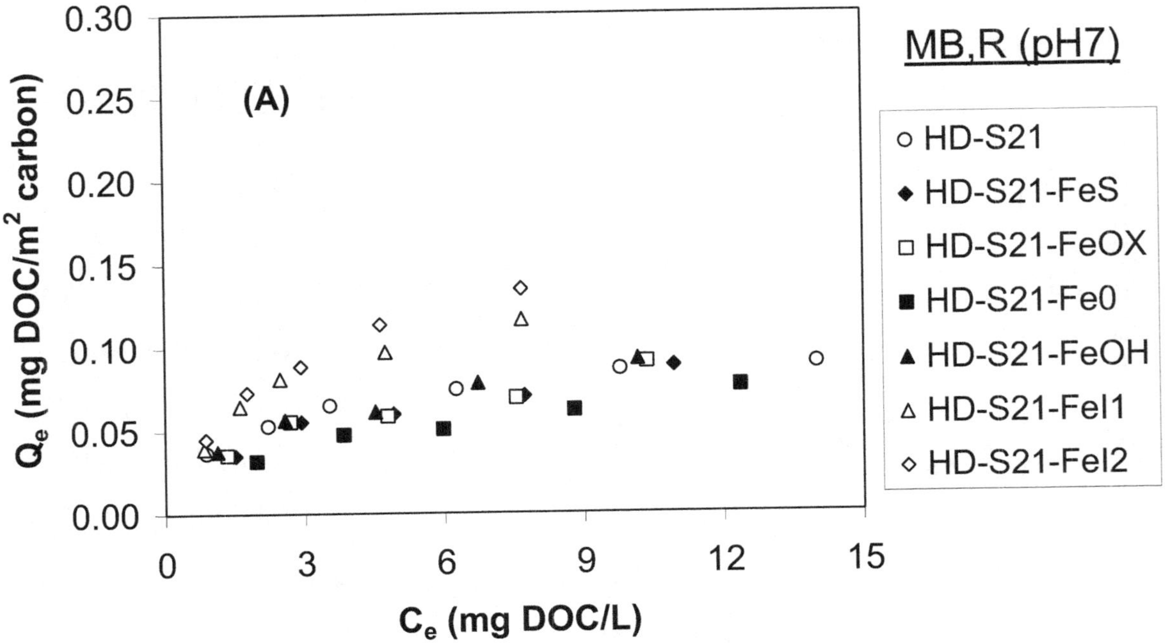

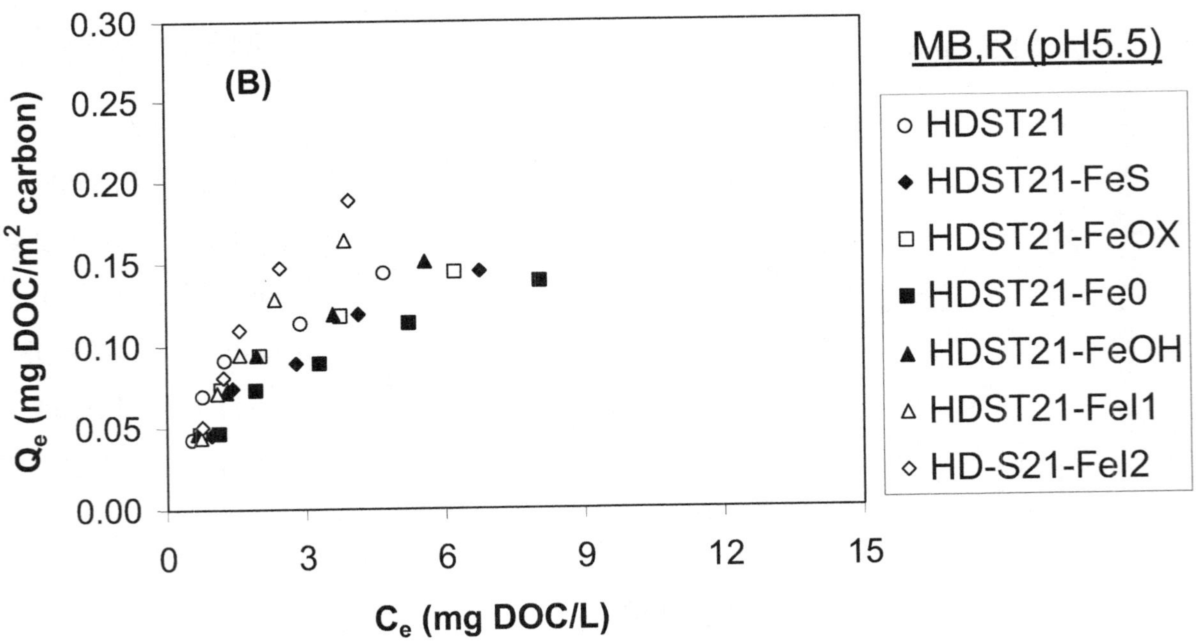

Figure 4.4 Surface-area basis Myrtle Beach raw water DOM adsorption isotherms of HD, HD-S21 and iron impregnated carbons at: (A) pH of 7, (B) pH of 5.5

Spartanburg treated water

Mass and surface-area based isotherms of Spartanburg treated water DOM at pH 5.5 and 7 by iron-impregnated carbons and their precursor, HD-S21, are shown in Figures 4.5 and 4.6, respectively. On a mass-basis, at equilibrium concentration of 3 mg TOC/L or higher, the DOM uptake of the carbons were in the order of HD-S21-FeI1 ≈ HD-S21-FeI2 > HD-S21 > HD-S21-FeOX > HD-S21-FeOH > HD-S21-FeS > HD-S21-Fe^0 at pH 7 and HD-S21-FeI1 ≈ HD-S21-FeI2 > HD-S21 ≈ HD-S21-FeOX ≥ HD-S21-FeOH > HD-S21-FeS > HD-S21-Fe^0 at pH 5.5. On a surface-area basis, the observed trends at both pH values appear to be in the order of HD-S21-FeI2 > HD-S21-FeI1 > HD-S21 ≈ HD-S21-FeOH ≈ HD-S21-FeOX > HD-S21-FeS ≈ HD-S21-Fe^0, grouped in three categories from the highest to the lowest uptake. As for the Myrtle Beach treated water, these trends did not correlate with the iron content of the carbons. However, for both treated waters, the most effective carbons were HD-S21-FeI1 and HD-S21-FeI2, which were obtained with the ion-exchange method from the initially oxidized precursor. Decreasing pH from 7 to 5.5 increased the DOM uptake of all carbons. The increase in the DOM uptake was 50%, 67%, 58%, 54%, 63%, 53% and 60% at C_e of 3 mg DOC/L for HD-S21, HD-S21-FeS, HD-S21-FeOH, HD-S21-FeOX, HD-S21-Fe^0, HD-S21-FeI1 and HD-S21-FeI2, respectively.

Spartanburg raw water

Mass based isotherms of Spartanburg raw water DOM at pH of 5.5 and 7 by iron-impregnated carbons and HD-S21 are shown in Figure 4.7. On a mass basis, the DOM uptake of iron-impregnated carbons showed different orders of uptake depending on pH. At pH of 7, for C_e of 3 mg TOC/L, the mass-basis uptake followed the order of HD-S21-FeI1 ≈ HD-S21-FeI2 ≈ HD-S21 ≈ HD-S21-FeOX ≥ HD-S21-FeOH > HD-S21-FeS > HD-S21-Fe^0 . At pH 5.5, the DOM uptake was in the order of HD-S21 > HD-S21-FeI1 ≈ HD-S21-FeI2 ≈ HD-S21-FeOX ≈ HD-S21-FeOH > HD-S21-FeS > HD-S21-Fe^0. In contrast to the Mrytle Beach raw water, the DOM uptakes of HD-S21-FeI1 and HD-S21-FeI2 carbons were not very different than the precursor carbon. This further indicates that the effectiveness of iron impregnated GACs will largely depend on specific interactions between some DOM components and iron species on the carbon surface.

When solution pH decreased from 7 to 5.5, the relative increase in the DOM uptake of HD-S21, HD-S21-FeS, HD-S21-FeOH, HD-S21-FeOX, HD-S21-Fe^0, HD-S21-FeI1 and HD-S21-FeI2 was 83%, 45%, 61%, 56%, 48%, 54%, and 56%, respectively, at C_e of 3 mg TOC/L. The extent of the increases in DOM uptake with decreasing pH was significantly lower than that of the Myrtle Beach raw water DOM. As discussed earlier, this can be attributed to the coiling effect with decreasing pH and more effective coiling of high than low molecular weight DOM components. The Myrtle Beach raw water DOM has much larger molecular weight and more hydrophobic character than the Spartanburg raw water DOM (Table 3.1).

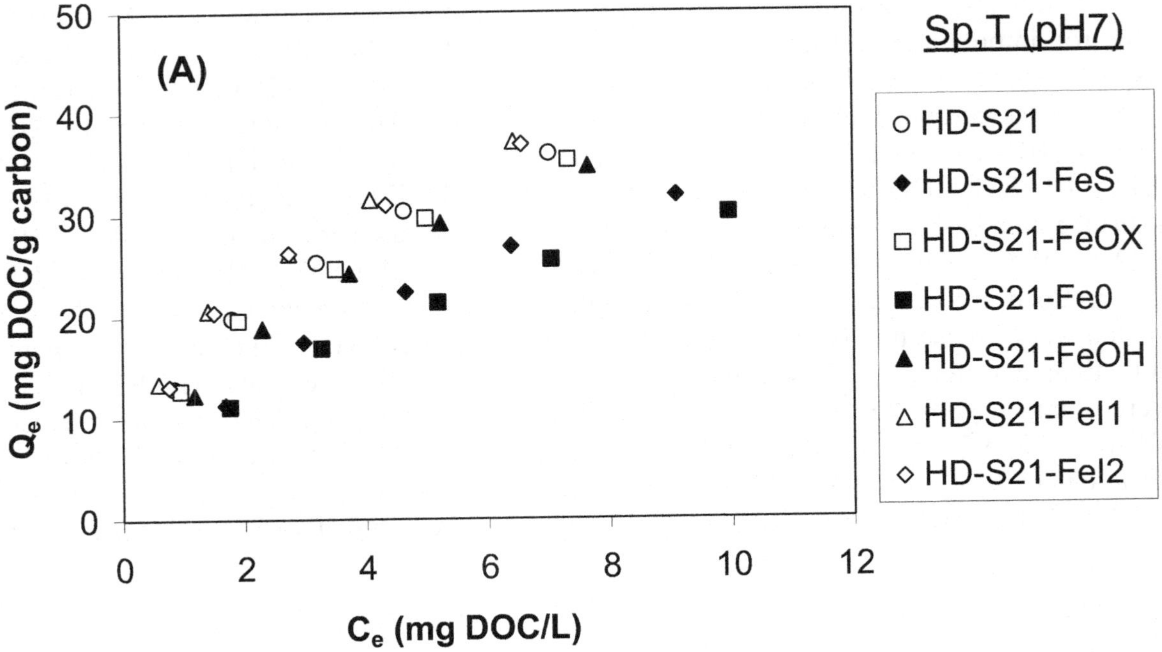

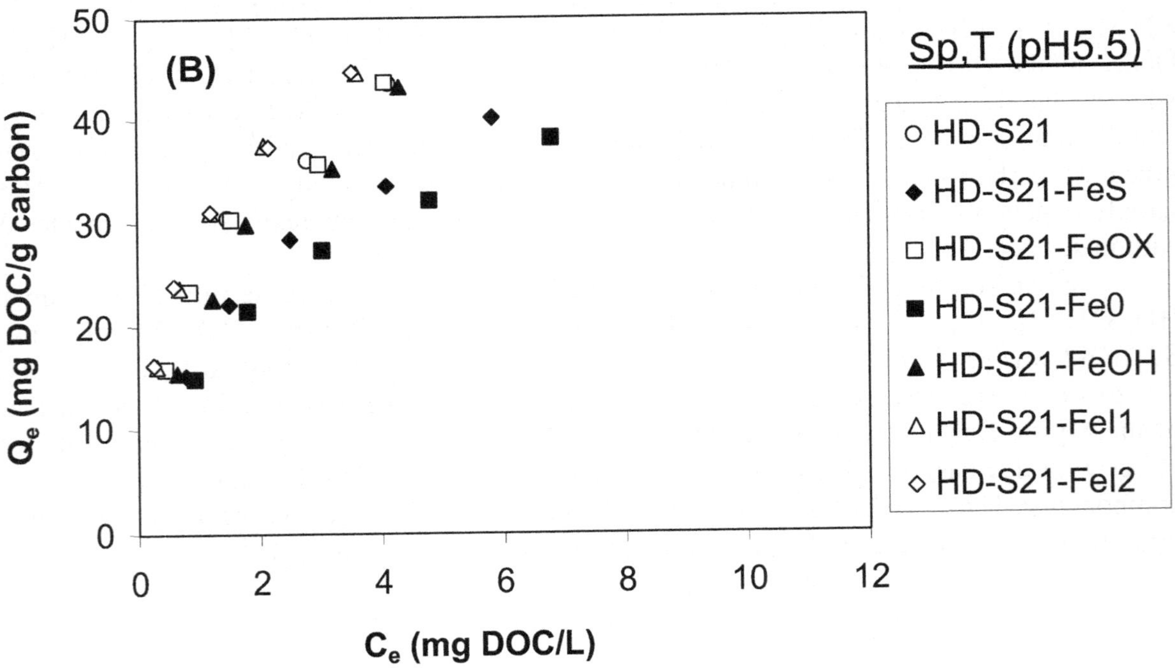

Figure 4.5 Mass basis isotherms of Spartanburg treated water DOM on HD-S21 and iron impregnated carbons at: (A) pH of 7 and (B) pH of 5.5

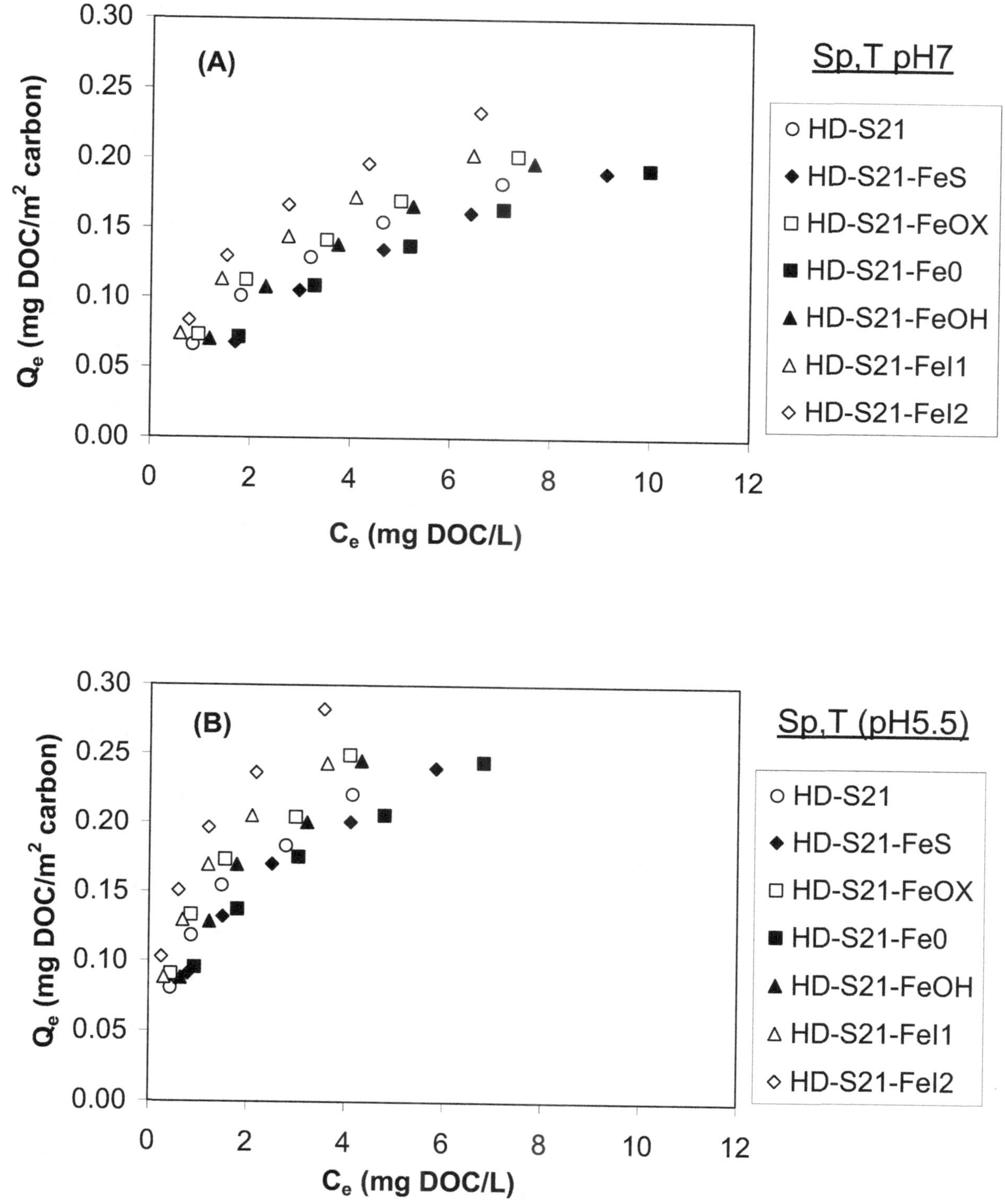

Figure 4.6 Surface-area basis isotherms of Spartanburg treated water DOM on HD-S21 and iron impregnated carbons at: (A) pH of 7, (B) pH of 5.5

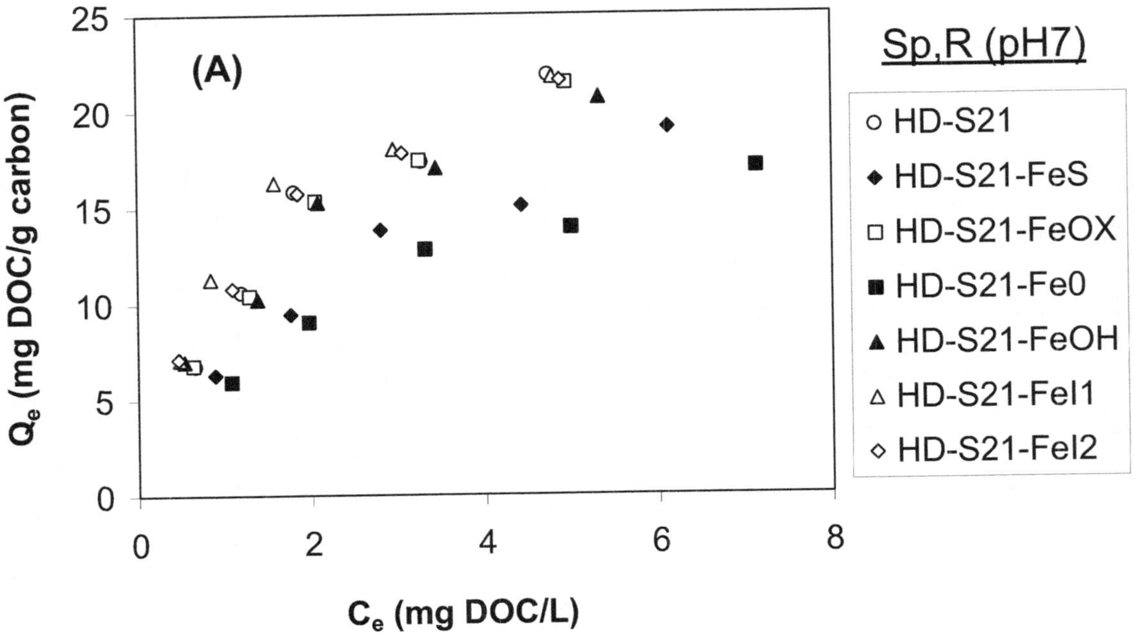

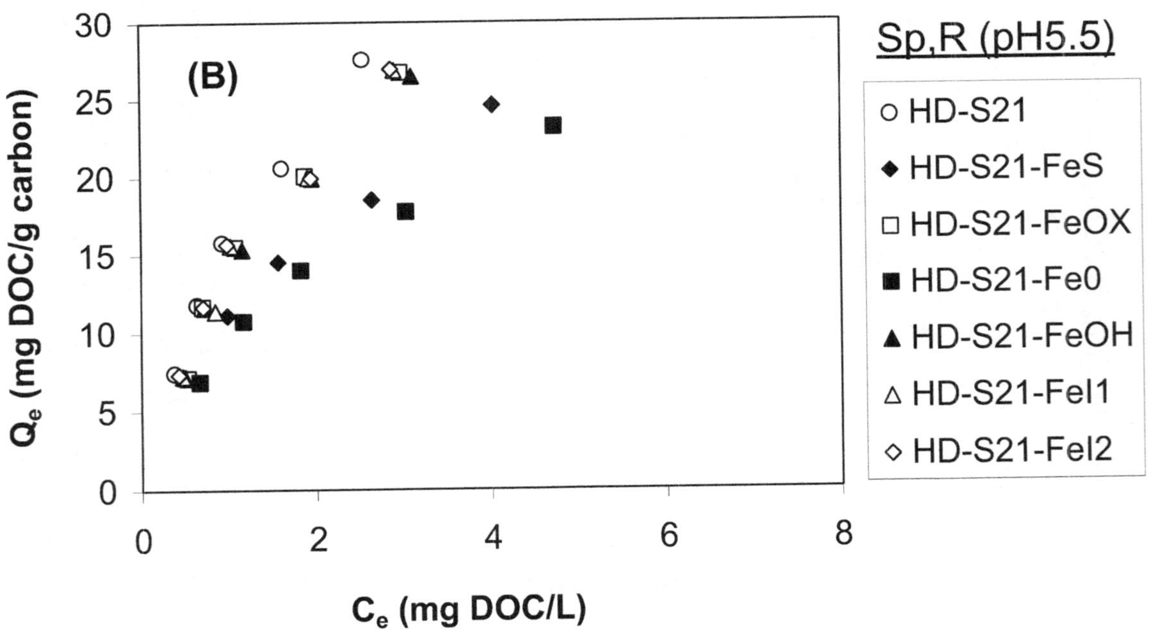

Figure 4.7 Mass basis Spartanburg raw water DOM isotherms of HD-S21 and iron-impregnated carbons at: (A) pH of 7 and (B) pH of 5.5

RSSCT Results

All the iron-impregnated carbons were tested on the Myrtle Beach treated water DOM at pH 6.5, and selected carbons were also tested at pH 5.5 and Spartanburg treated water DOM at both pHs. RSSCT experiments employed treated waters because most water utilities use GAC fixed bed filters after conventional treatment. The RSSCT breakthrough curves are presented as a function of both bed volume of water treated and large scale (EBCT=10 min) column operation time.

Myrtle Beach treated water

The RSSCTs results of the Myrtle Beach treated water on the two best performing iron-impregnated carbons, HD-S21-FeI1 and HD-S21-FeI2, in the isotherm tests are presented in Figures 4.8 to 4.12. The TOC and UV_{254} breakthrough profiles for HD-S21-FeI1 (Figure 4.8) and HD-S21-FeI2 (Figure 4.11) indicated that there was no additional DOM removal by iron-impregnated carbons as compared to their precursor HD-S21. This was not in agreement with the isotherm results, where HD-S21-FeI1 and HD-S21-FeI2 showed higher uptake than the precursor HD-S21. The column effluent samples were tested for iron leaching, and iron concentrations was always below the detection limits, indicating that iron was stable on the carbon surface during the RSSCTs. One possibility for the lack of iron effect during the column experiments may be related to the differences in the adsorption reactivity and kinetics of different DOM components between batch isotherm and column experiments. It is likely that there were some very reactive DOM components in the water that can adsorb very fast on both the carbon backbone and on iron groups. Since there was a continuous supply of these reactive species in the influent water of the RSSCT, they could outcompete and/or displace other DOM species which had higher affinity toward iron sites, but may have relatively slower sorption kinetics. In an isotherm bottle, there was a limited supply of reactive DOM components. Once these species were adsorbed, the remaining DOM species that have a tendency to adsorb, especially to iron sites, can also be removed from solution, because there was enough time (2 weeks) and no more competition from the other species for their adsorption.

The results of %HPO, $SUVA_{254}$ breakthrough profiles and DBP formation data were in general consistent with each other and with the trends of TOC and UV_{254} breakthrough profiles (Figures 4.9, 4.10 and 4.12). It is evident that iron impregnation did not show any benefit for either DOM removal or DBP control as compared to the precursor carbon, HD-S21. These results further emphasize the necessity of performing RSSCTs, despite their more costly and time demanding nature as compared to batch isotherm experiments, to obtain practically relevant information to evaluate or screen different adsorbents.

The TOC breakthrough profiles of the other iron-impregnated carbons, HD-S21-FeS, HD-S21-FeOH, HD-S21-FeOX and HD-S21-Fe^0, are plotted in Figures 4.13 to 4.16. Their UV_{254}, HPO%, $SUVA_{254}$ and DBP formation breakthrough profiles are provided in Appendix D. Similar to the isotherm results, these carbons did not show any iron-impregnation benefit on the DOM removal and DBP formation reduction.

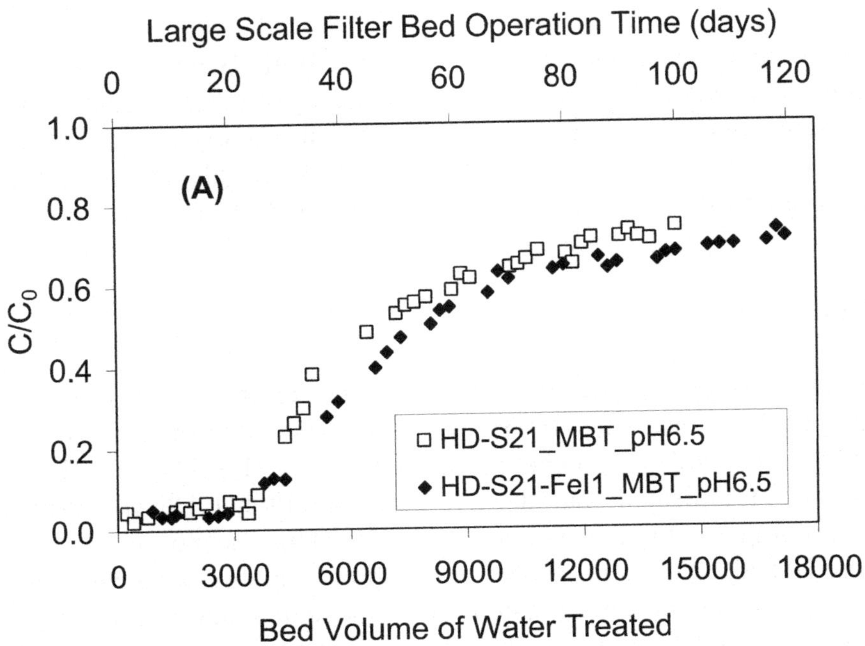

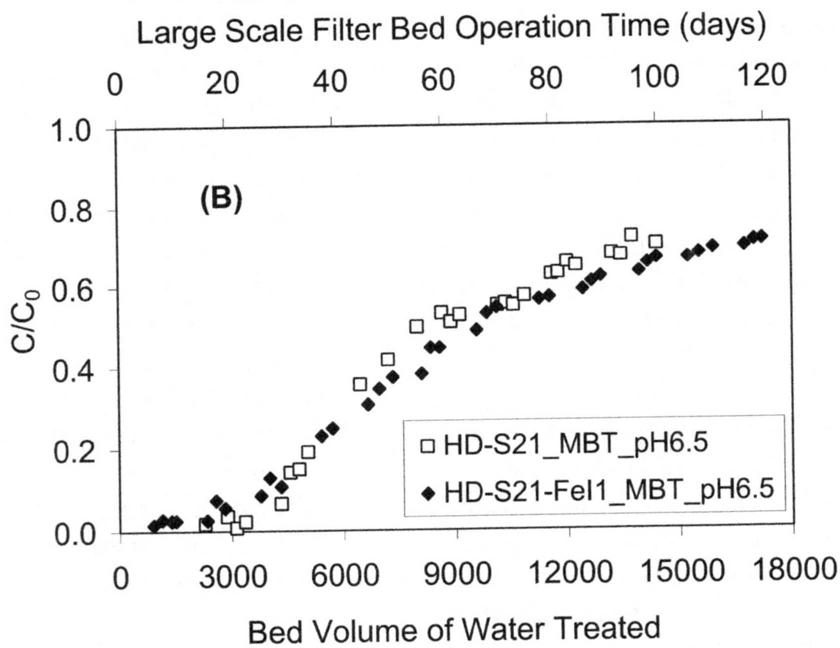

Figure 4.8 RSSCT results, expressed as (A) TOC, (B) UV_{254}, obtained for HD-S21 and HD-S21-FeI1 carbons with Myrtle Beach treated water DOM at pH of 6.5

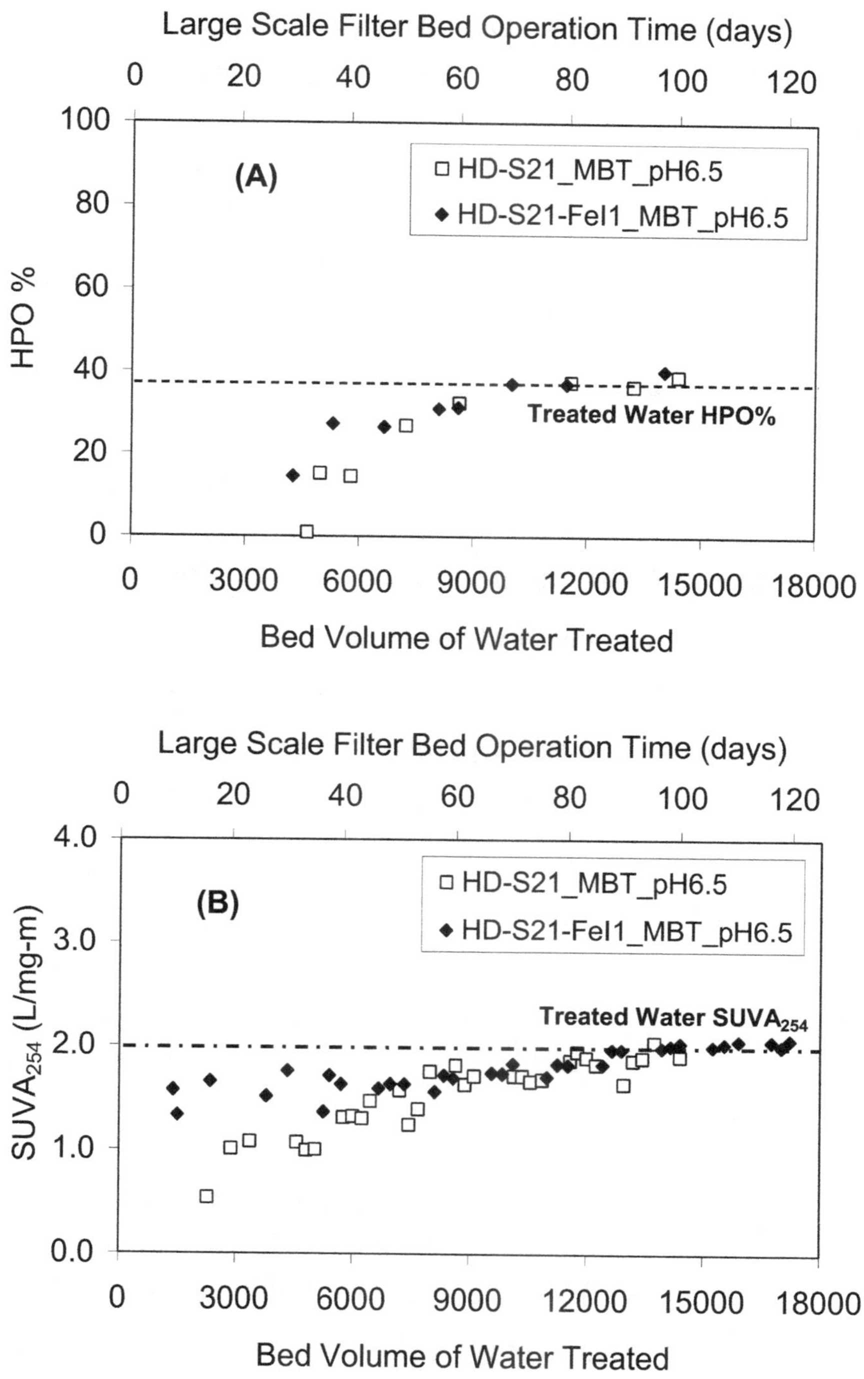

Figure 4.9 Myrtle Beach treated water DOM (A) HPO% and (B) SUVA$_{254}$ breakthrough profiles for HD-S21-FeI1

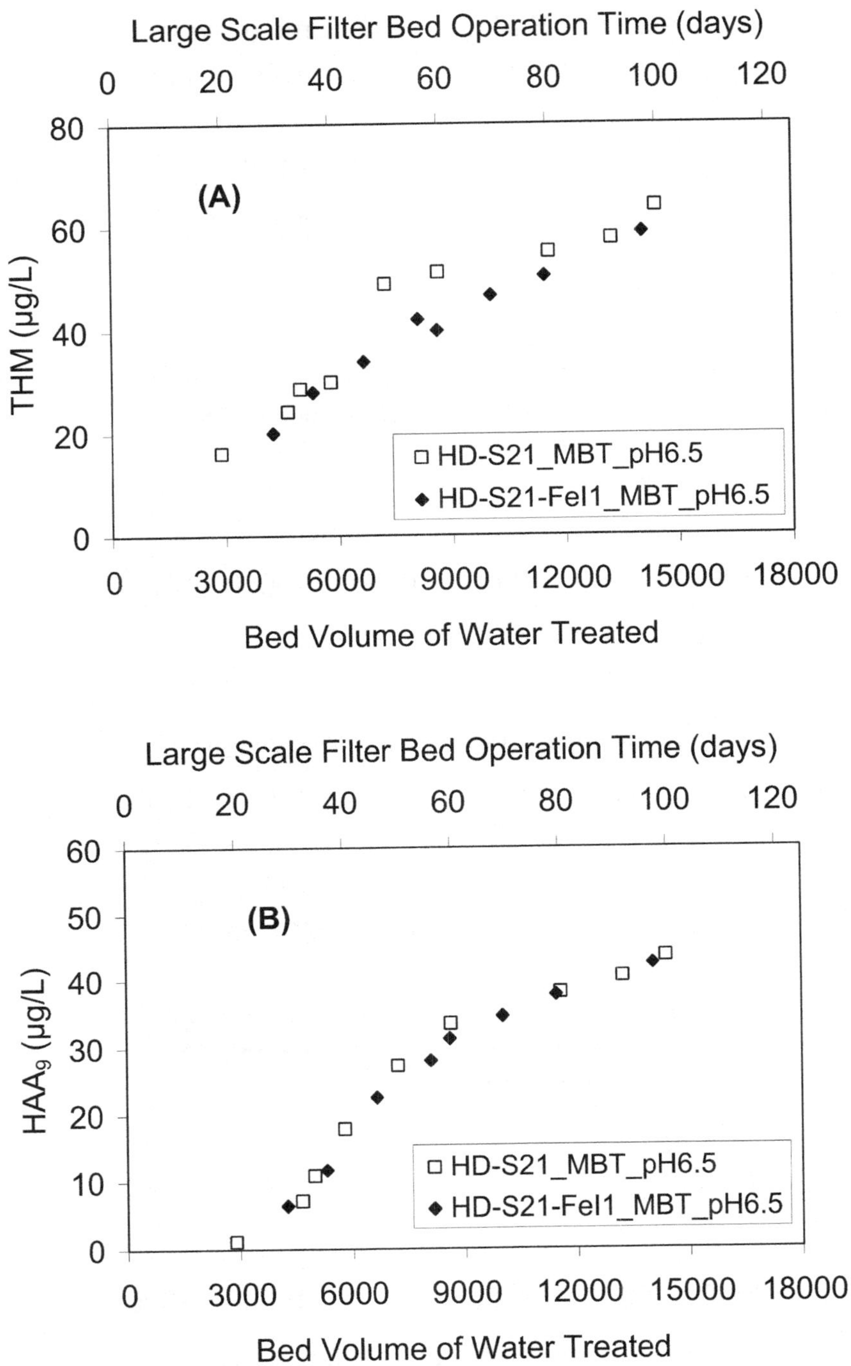

Figure 4.10 Myrtle Beach treated water DOM (A) THM and (B) HAA_9 formation breakthrough profiles for HD-S21 and HD-S21-FeI1

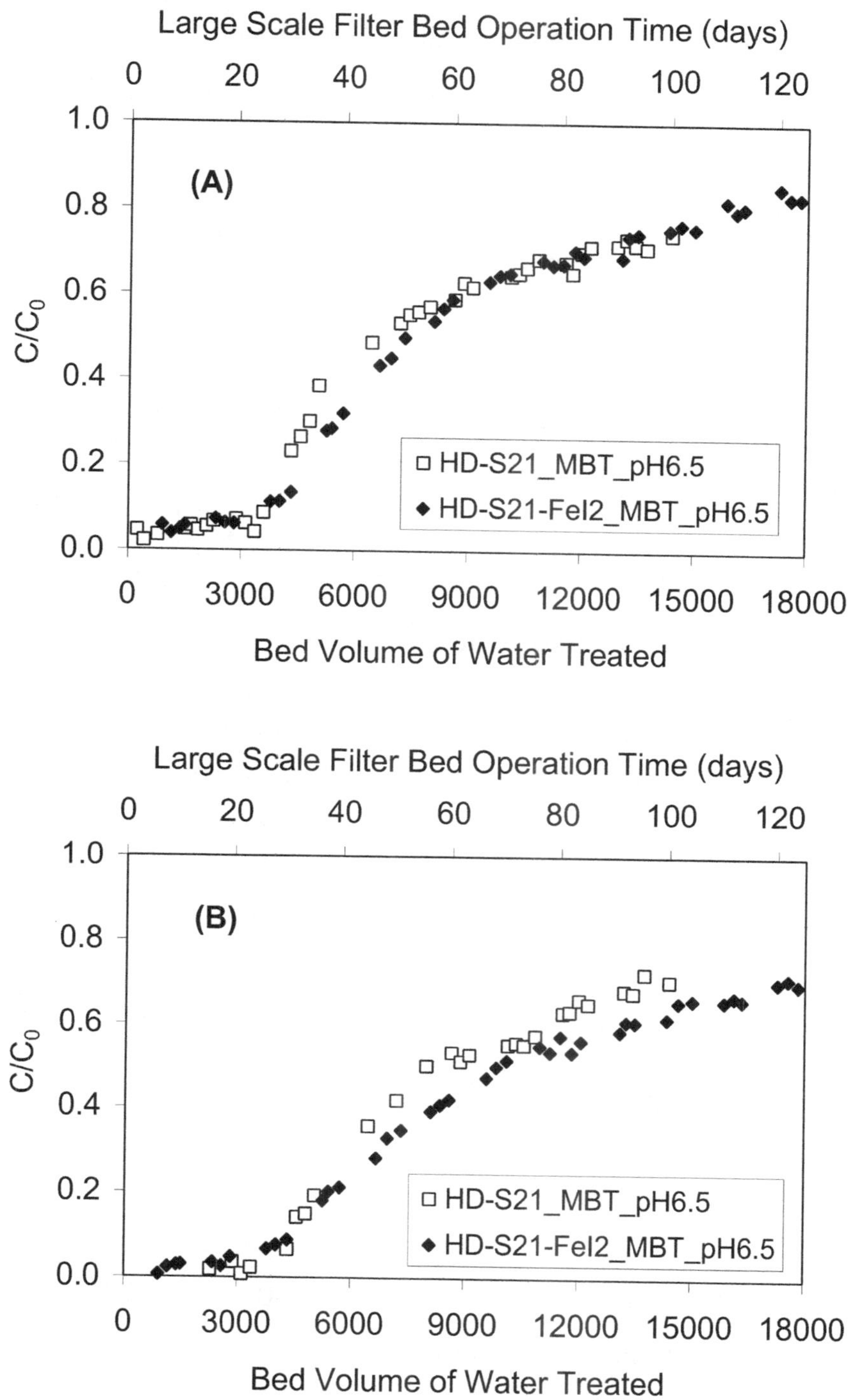

Figure 4.11 RSSCT results, expressed as (A) TOC, (B) UV_{254}, obtained for HD-S21 and HD-S21-FeI2 carbons with Myrtle Beach treated water DOM at pH of 6.5

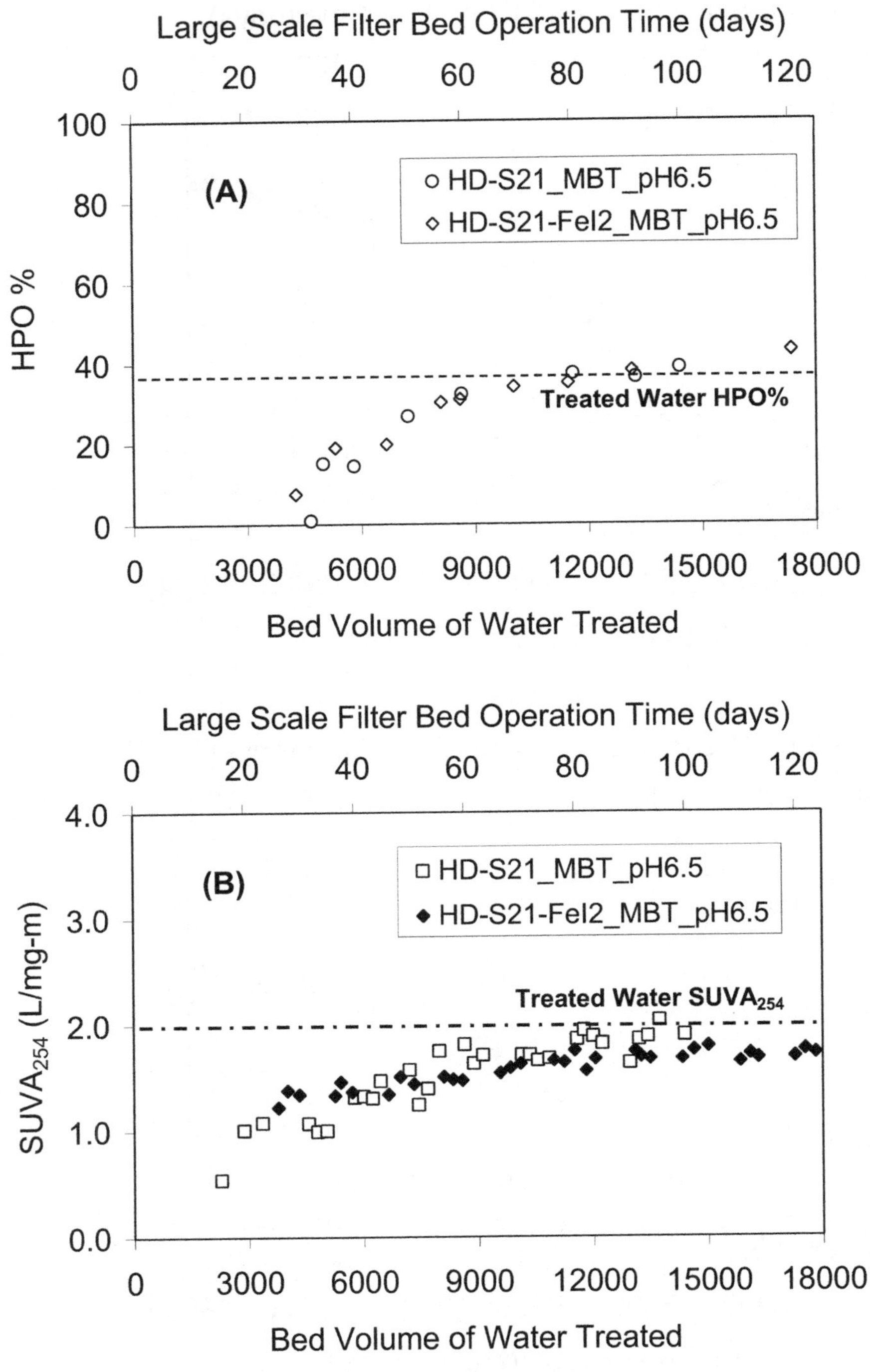

Figure 4.12 Myrtle Beach treated water DOM (A) HPO% and (B) $SUVA_{254}$ breakthrough profiles for HD-S21-FeI2

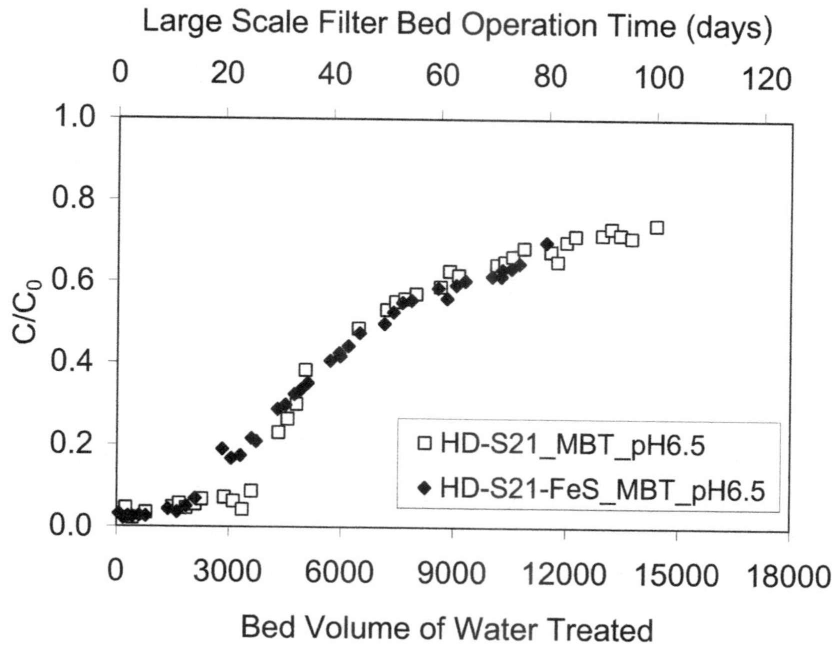

Figure 4.13 RSSCT results, expressed as TOC, obtained for HD-S21 and HD-S21-FeS carbons with Myrtle Beach treated water DOM at pH of 6.5

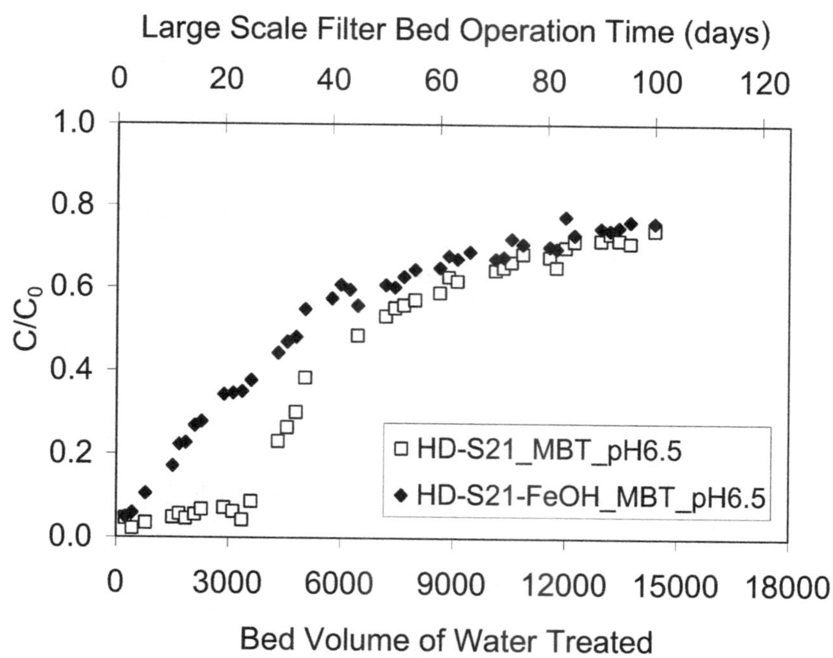

Figure 4.14 RSSCT results, expressed as TOC, obtained for HD-S21 and HD-S21-FeOH carbons with Myrtle Beach treated water DOM at pH of 6.5

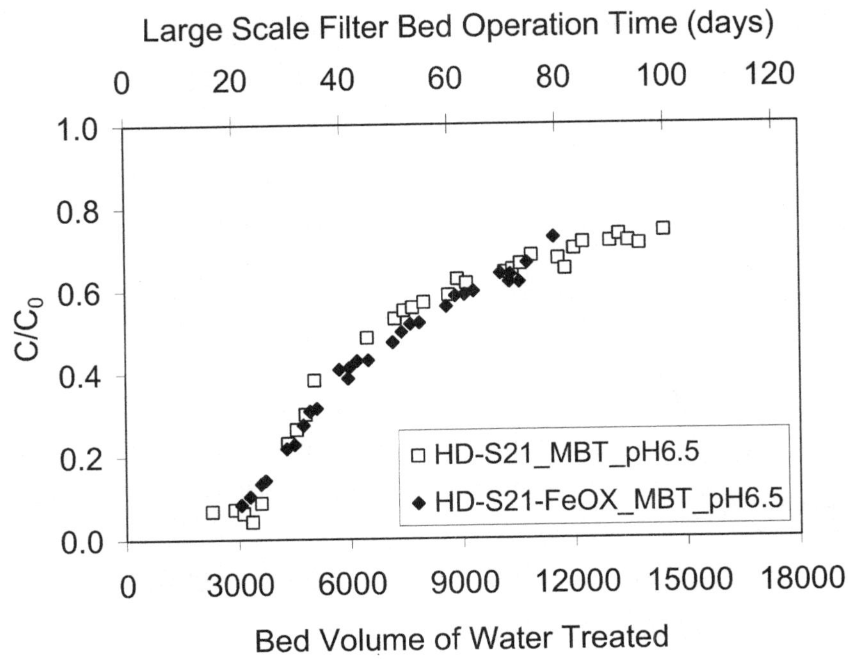

Figure 4.15 RSSCT results, expressed as TOC, obtained for HD-S21 and HD-S21-FeOX carbons with Myrtle Beach treated water DOM at pH of 6.5

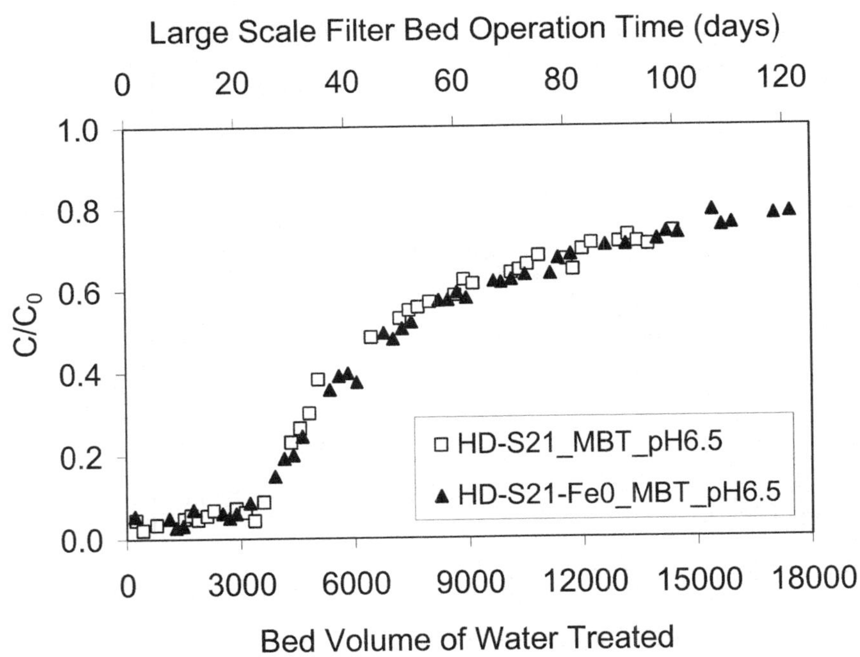

Figure 4.16 RSSCT results, expressed as TOC, obtained for HD-S21 and HD-S21-Fe^0 carbons with Myrtle Beach treated water DOM at pH of 6.5

Two iron-impregnated GACs, HD-S21-FeI1 and HD-S21-FeOX, and their precursor HD-S21, were also tested with the Myrtle Beach treated water at pH 5.5 to investigate the pH effect on DOM removal. The comparison of the TOC breakthrough profiles at pH 6.5 and 5.5 are plotted in Figures 4.17 and 4.18. The UV_{254}, HPO%, $SUVA_{254}$ and DBP formation breakthrough profiles of HD-S21-FeI1 and HD-S21-FeOX are in Appendix D, respectively. The results showed that there was a slight increase of operation time (or bed volume of water treated) when pH decreased from 6.5 to 5.5. However, the iron-impregnated carbons still performed similarly to their precursor, HD-S21. Again, no iron effect on DOM removal was observed at the lower pH value. As a result, there was no obvious effect on the THM and HAA_9 formation reduction in the effluents of RSSCTs (Appendix D).

Spartanburg treated water

To investigate whether there was an iron effect with the hydrophilic Spartanburg water DOM, RSSCTs were run using one of the best performing iron-impregnated carbons, HD-S21-FeI1, at both pH 6.5 and 5.5. The breakthrough profiles are shown in Figures 4.19 to 4.21.

It is noted that at pH 6.5, the HD-S21-FeI1 breakthrough was faster than that of HD-S21 in both TOC and UV_{254} profiles. However, this was not observed at pH 5.5. This difference was also seen in the THM breakthrough profiles (Figure 4.21) but not for HAA_9. The HPO%, $SUVA_{254}$ and THM formation breakthrough profiles also captured this difference (Figures 4.20 to 4.21). The cause of the different breakthrough patterns observed at pH 6.5 is not known. However, it is evident that no clear advantage of iron impregnation was observed on the reduction of DBP formation. Decreasing pH increased the removal of both THM and HAA_9 precursors for HD-S21-FeI1. It was also noted that, in all the breakthrough profiles for the Myrtle Beach and Spartanburg waters, UV_{254} breakthrough was always delayed as compared to TOC breakthrough. This indicated that, as for steam-treated carbons, UV_{254} absorbing components were preferentially removed from water by iron impregnated carbons.

An RSSCT experiment was also performed for HD-S21-FeOX GAC. No clear evidence of any iron effect was observed (Figure 4.22). This was also supported by the HPO%, $SUVA_{254}$, THM and HAA_9 formation breakthrough profiles (Appendix D).

Comparison of DOM uptake during isotherm and RSSCT

The amounts of DOM uptake during the batch isotherm and RSSCT experiments with both Myrtle Beach and Spartanburg waters by the iron-impregnated GACs and their precursor were calculated in Table 4.2. Generally, the DOM uptake from column experiments, although it was only up to 70% of the breakthrough, was significantly higher than the uptake obtained from the isotherms. The possible reasons have been provided in the discussion of steam-treated carbons RSSCT, and include the heterogeneous nature of DOM and different adsorption patterns between column and isotherm experiments (Chapter 3). Therefore, as discussed before, it is likely that there were some very reactive DOM components in the natural waters that can adsorb very quickly on both carbon backbone and to iron groups. Since there was a continuous supply of these reactive species in the influent water of RSSCT, they could outcompete other DOM species which had higher affinity toward iron sites, but may had slower sorption kinetics. In an isotherm bottle, there was a limited supply of reactive DOM components. Once these species were adsorbed, the remaining DOM species that have a tendency to adsorb, especially to iron sites, can also be removed from solution, because there was enough time (2 weeks) and no competition from the other species for their adsorption.

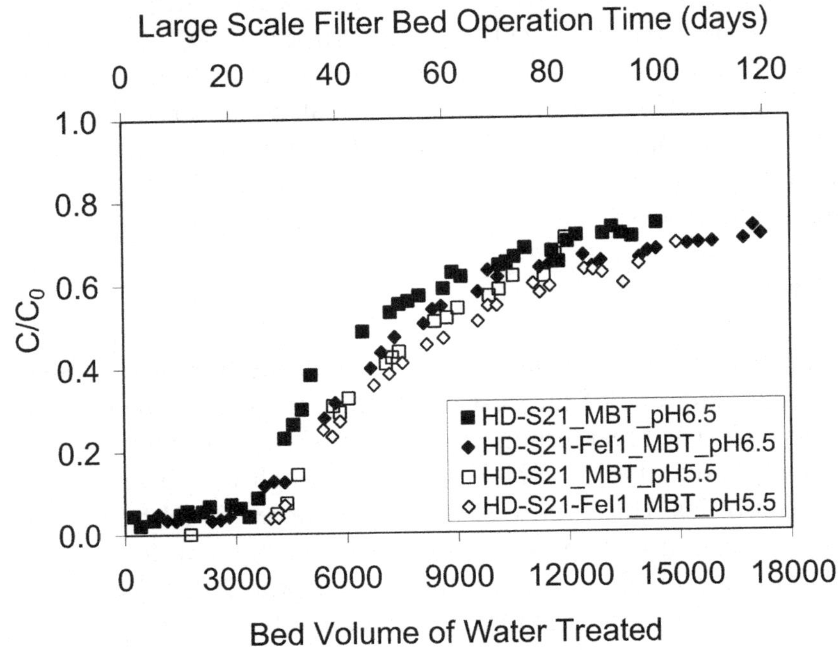

Figure 4.17 RSSCT results, expressed as TOC, obtained for HD-S21 and HD-S21-FeI1 carbon with Myrtle Beach treated water DOM at pH 6.5 and 5.5

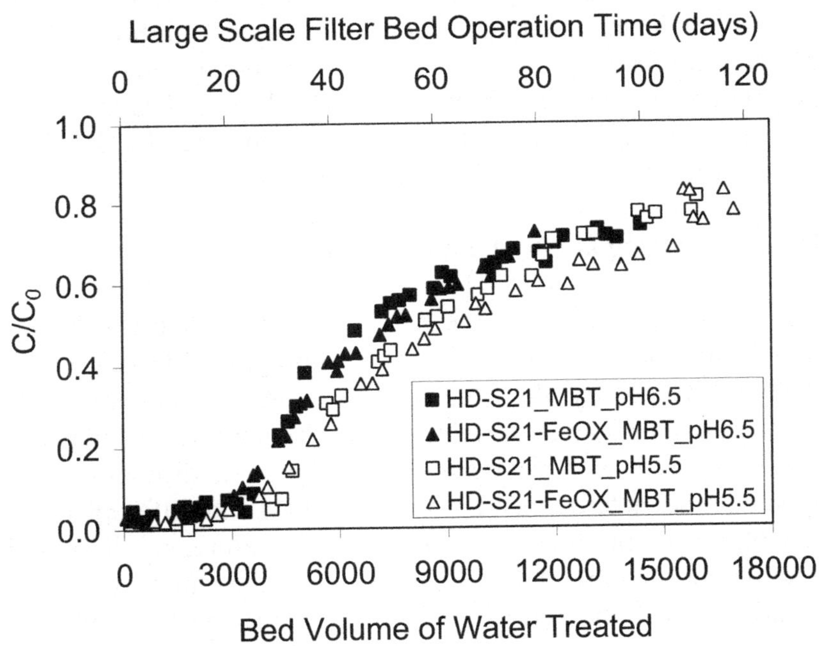

Figure 4.18 RSSCT results, expressed as TOC, obtained for HD-S21 and HD-S21-FeOX carbon with Myrtle Beach treated water DOM at pH 6.5 and 5.5

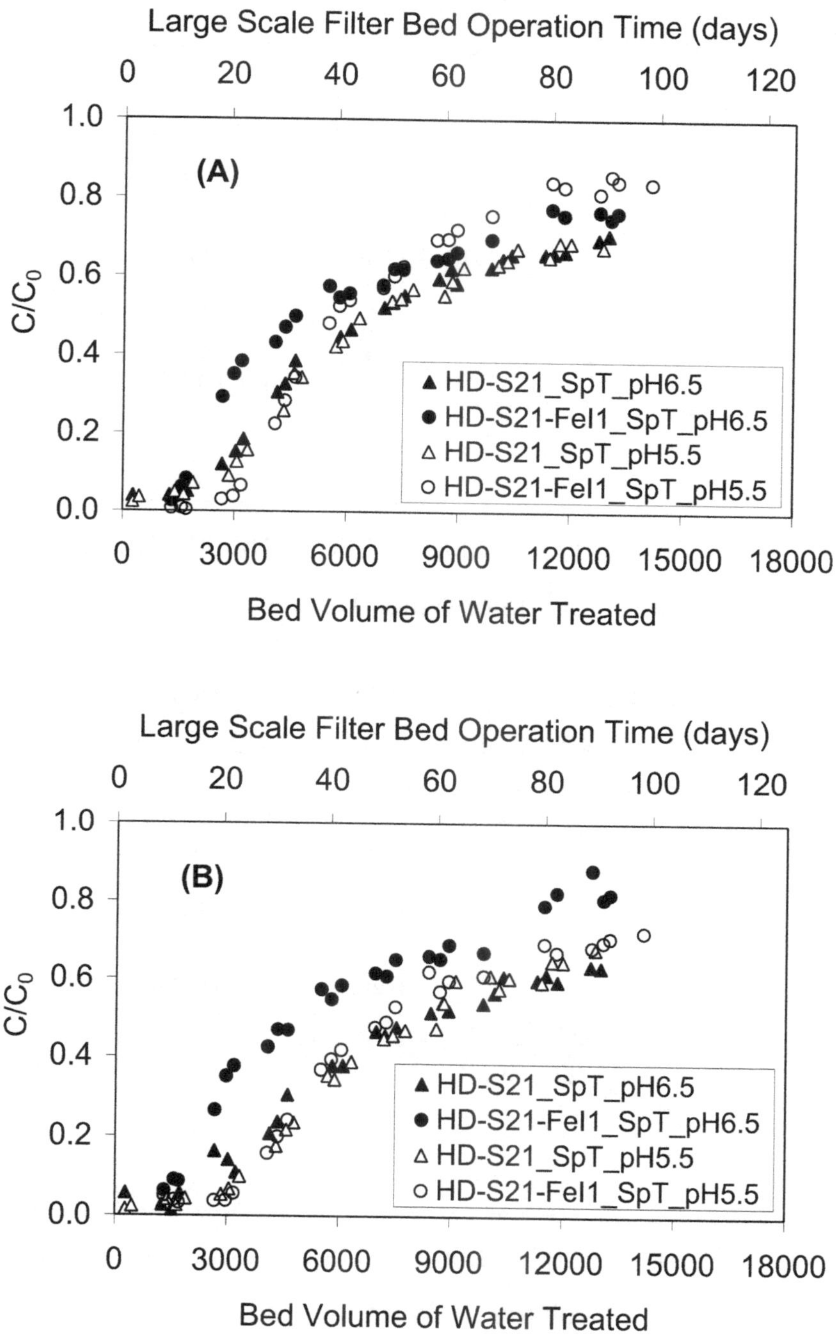

Figure 4.19 RSSCT results, expressed as (A) TOC and (B) UV_{254}, obtained for HD-S21 and HD-S21-FeI1 carbon with Spartanburg treated water DOM at pH 6.5 and 5.5

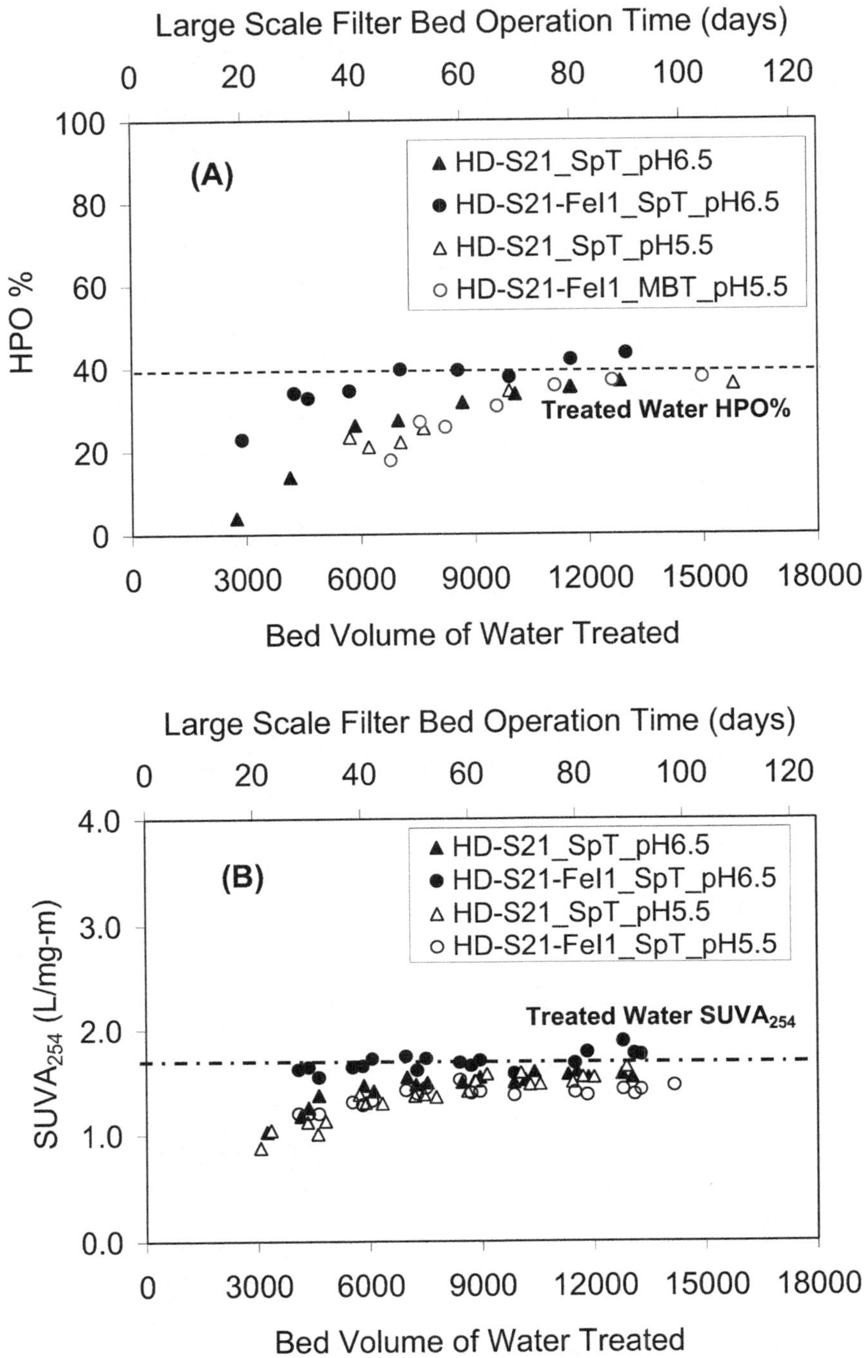

Figure 4.20 Spartanburg treated water DOM (A) HPO% and (B) $SUVA_{254}$ breakthrough profiles for HD-S21 and HD-S21-FeI1

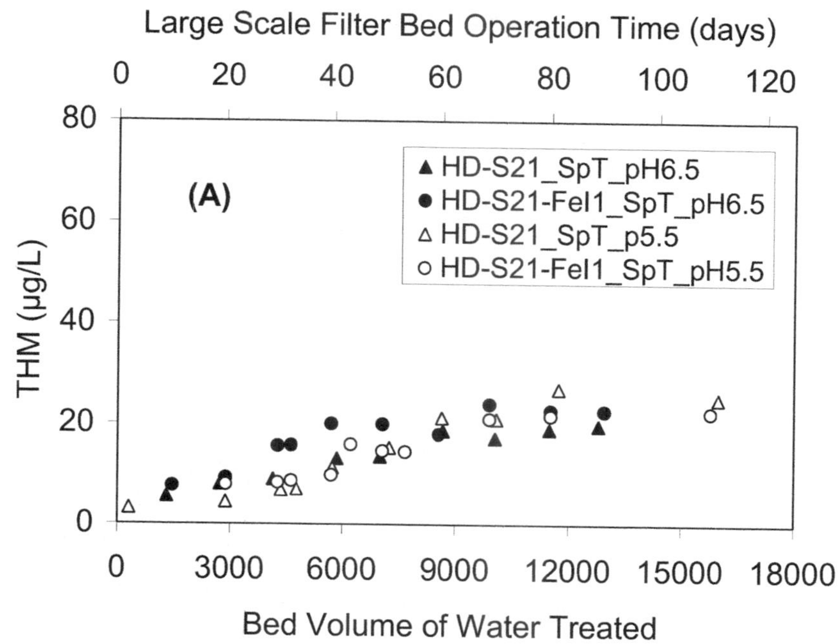

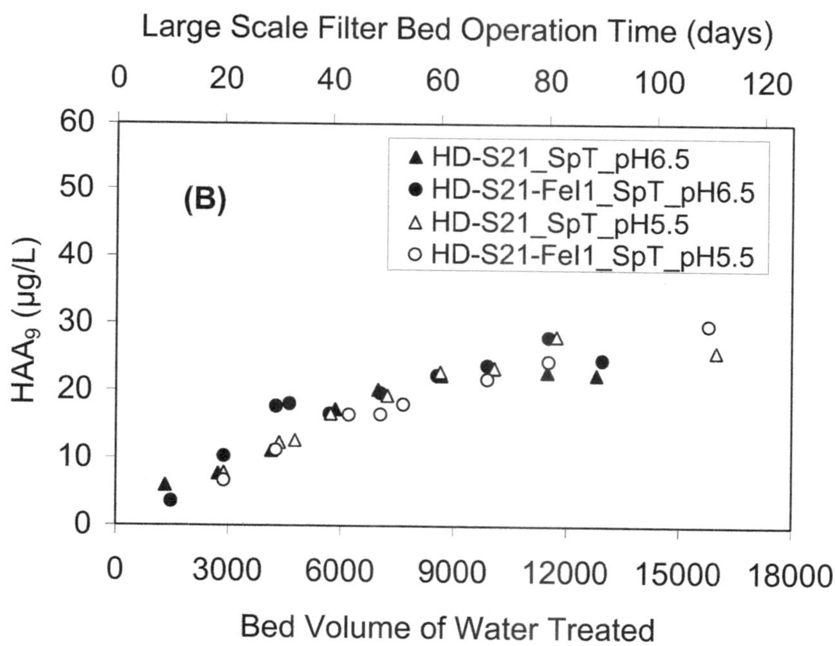

Figure 4.21 Spartanburg treated water DOM (A) THM and (B) HAA_9 formation breakthrough profiles for HD-S21 and HD-S21-FeI1 at pH 6.5 and 5.5

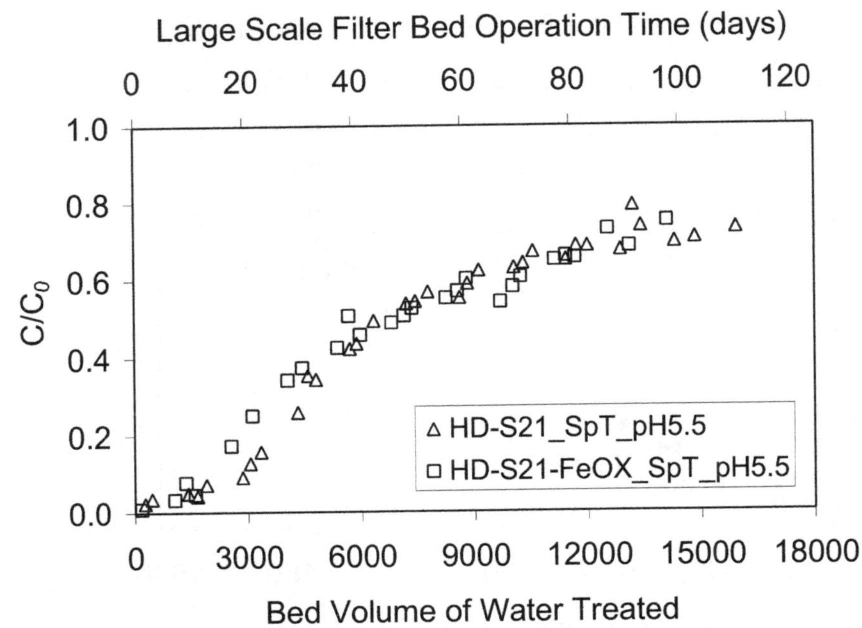

Figure 4.22 Spartanburg treated water DOM (TOC) breakthrough profiles for HD-S21 and HD-S21-FeOX at pH 5.5

Table 4.2
Bed volume of water treated and DOM uptake by the virgin and iron-impregnated carbons for Myrtle Beach and Spartanburg treated waters until 70% of TOC breakthrough, and the DOM uptake at C_e=3 mg/L from the isotherm experiments

Water Source	pH	Carbon	Up to 70% Breakthrough		Isotherm at Ce=3 mg/L
			Bed Volume of Water Treated	DOM uptake mg TOC/g GAC	DOM uptake mg TOC/g GAC
MBT	6.5	HD	8580	31.8	18.8
		HD-S21	12000	44.9	23.6
		HD-S21-FeS	11460	36.0	20.0
		HD-S21-FeOX	11460	31.5	20.0
		HD-S21-Fe^0	12000	30.5	16.9
		HD-S21-FeOH	10860	23.9	11.7
		HD-S21-FeI1	16750	57.9	31.3
		HD-S21-FeI2	13095	49.4	31.3
	5.5	HD	11340	43.5	23.8
		HD-S21	15960	60.4	29.3
		HD-S21-FeI1	14940	46.9	38.8
		HD-S21-FeOX	15300	55.5	24.7
SpT	6.5	HD	9120	30.6	19.8
		HD-S21	13020	44.6	24.6
		HD-S21-FeI1	9870	34.6	26.8
	5.5	HD	8820	32.0	29.4
		HD-S21	14850	51.6	37.4
		HD-S21-FeI1	8700	30.0	40.7
		HD-S21-FeOX	12540	39.5	35.8

SUMMARY OF IMPORTANT FINDINGS

Iron impregnation of activated carbons may increase the DOM uptake. However, the effectiveness of iron impregnation depended on the preparation technique employed. The results indicated that (i) iron loading, (ii) iron dispersion on the carbon surface, (iii) iron species formed, (iv) the impact of iron deposition on carbon pore size distribution, (v) the final surface charge of the carbon surface, and (vi) the DOM composition would all impact the DOM removal by iron-impregnated activated carbons

Two iron-impregnated GACs prepared by ion-exchange methods, HD-S21-FeI1 and HD-S21-FeI2, generally exhibited the highest DOM removal in the isotherm tests. However, the degree of iron effect depended on the DOM components. A higher degree of iron effect was observed with the high $SUVA_{254}$ than low $SUVA_{254}$ DOM.

Although the isotherm results showed that some iron-impregnation of the activated carbons could increase DOM removal, no similar effect was observed when the same carbon was tested in the RSSCTs. This further illustrated the complexity of enhanced DOM removal and DBP control using iron-impregnated GACs. Therefore, it was concluded that it would be practically more feasible and effective to prepare steam-treated or mesoporous activated carbons than iron-impregnated GACs for DOM and DBP control. The results both in this and in the previous chapters have also clearly indicated that it is essential to conduct a RSSCT, despite its longer and more costly nature, experiment to evaluate the effectiveness of different carbons for DOM removal.

CHAPTER 5
CONCLUSIONS AND RECOMMENDATIONS

CONCLUSIONS

Important findings from this project are as follows:

- The results obtained in this study using both raw and treated water DOM at the typical pH range of water treatment (i.e., pH 5.5 to 8) provided direct experimental evidence that the great majority of DOM components in natural waters cannot access pores smaller than 1 nm. Therefore, when GAC adsorption is intended to use for DOM removal, GACs with large surface areas and volumes in pores >1 nm should be selected. BET surface area and total pore volume are not the important parameters in selecting GAC for DOM removal because they do not provide information about the *accessible* pore region. GAC surface chemistry also plays a role in DOM adsorption. It was shown that high surface acidity (i.e., acidic pH_{PZC} values) significantly decreases DOM adsorption. However, the effect of pore size distribution becomes more important than pH_{PZC} at $pH_{PZC}>7$.
- GACs with larger pore sizes and higher pH_{PZC} can be produced by extending steam treatment periods during carbon preparation. The steam-treated GACs showed higher DOM uptakes compared with their precursor. However, the increase in DOM uptake was not always proportional to the extent of steam treatment. The increase in DOM uptake was mainly due to the removal of DOM components with more hydrophobic character and higher $SUVA_{254}$ values.
- The physicochemical characteristics of DOM play important roles in adsorption. Enlarging carbon pores by steam treatment enhanced the DOM uptake; however extended steam treatment period and additional pore enlargement did not always result in additional DOM adsorption for all waters. Although the higher accessible pore volume and higher pH_{PZC} as a result of steam treatment resulted in higher uptake of the Myrtle Beach treated water DOM by HD-S45 than HD-S21, the difference between the Spartanburg treated water DOM adsorption by HD-S45 and HD-S21 was negligible. For both waters, HD-S21 exhibited significantly higher DOM adsorption than the virgin HD. Myrtle Beach and Spartanburg treated DOM had similar $SUVA_{254}$ and MW_w values. The increase in the DOM uptake as a result of steam treatment was not found to correlate with simple DOM characterization parameters, such as $SUVA_{254}$ and MW_w determined through size exclusion chromatography. It appears that additional differences (e.g., the *size* and structure) in the DOM mixtures that were not fully resolved by $SUVA_{254}$ and MW_w measurements caused differences in their adsorption by activated carbons.
- Removal of high molecular weight DOM components from water with conventional treatment processes (coagulation/flocculation/sedimentation) significantly increases the operational life of GAC filters. Therefore, the impact of DOM removal during conventional treatment processes on the performance of subsequent GAC adsorption for DBP formation control should be evaluated.
- The RSSCT results for Myrtle Beach and Spartanburg water DOM at pH 6.5 and 5.5 indicated that: (1) Steam-treated carbons had a longer operation time than their precursor to reach a certain degree of breakthrough, (2) pH effect on the DOM removal depends on the characteristics of source water. For the high $SUVA_{254}$ and more hydrophobic Myrtle Beach

water, decrease in pH increased the DOM uptake significantly, whereas for the low $SUVA_{254}$ and more hydrophilic Spartanburg water, the pH effect was minimal, (3) steam-treated GACs can show higher degree of THM and/or HAA_9 precursor removal. However, the extent of removal was water specific and pH dependent. In this study, for the Myrtle Beach treated water, after 18,000 bed volumes or approximately 4+ months of operation, the effluent of steam-treated HD-S45 (at pH 5.5) column had THM and HAA_9 formation well below the regulatory limits, 80 and 60 ppb respectively, and (4) UV absorbing DOM molecules were preferentially removed by the steam-treated GACs.

- From a DBP control perspective, the effectiveness of GAC adsorption should be evaluated by performing DBP formation tests. Although, from a practical point of view, batch isotherm measurement is a simple technique that can be used to screen various activated carbons for DOM removal, a dynamic RSSCT, despite its time consuming and costly experimental nature, should be performed to obtain practically relevant information about the feasibility of using activated carbon adsorption for a full-scale application. The DBP formation in the effluent of RSSCT should be tested because monitoring of TOC or UV_{254} removal alone is not going to provide conclusive information about the effectiveness of GAC adsorption for DBP control.
- Regeneration experiments showed that high temperature treatment under H_2 or N_2 atmospheres effectively restored the DOM adsorption capacity of steam-treated GACs for at least two cycles of regeneration. In contrast, the NaOH regenerated carbons showed a continuous trend of decreasing DOM uptake with increasing regeneration cycle.
- Among several iron impregnation methods, ion exchange showed the most enhanced DOM at typical water treatment conditions based on isotherm experiments. However, the impact of iron impregnation also depended on the DOM characteristics. Isotherm results showed higher DOM uptakes of both Myrtle Beach treated and raw waters by iron-impregnated carbons with the ion exchange method. However, the effect of iron impregnation was minimal for the adsorption of Spartanburg raw and treated water DOM. Since both Myrtle Beach and Spartanburg treated waters have similar $SUVA_{254}$ and MW_w values, it appears that there were some other characteristics of DOM that were responsible for the higher affinity of Myrtle Beach water DOM toward the iron-impregnated GACs.
- The advantage demonstrated by iron-impregnated carbons prepared using the ion-exchange method during batch isotherm tests was not observed in RSSCTs. The lack of iron impact was attributed to the different adsorption patterns between the column and batch mode isotherm tests. It appears that there are some very reactive DOM components in the water that can adsorb very fast on both carbon backbone and iron sites. Since there is a continuous supply of these reactive species in the influent of RSSCT, they can outcompete other DOM species which have higher affinity toward iron sites but probably with relatively slower sorption kinetics. On the other hand, in an isotherm bottle, there is a limited supply of reactive DOM components. Once these species adsorb, the remaining DOM species that have tendency to adsorb, especially to iron sites, can also be removed from solution, because there is enough time and no more competition for their adsorption. This explanation is also in agreement with the higher DOM adsorption capacities observed during RSSCT as compared to the isotherm experiments.

RECOMMENDATIONS TO WATER UTILITIES

Complying with the stringent DBP regulations require water utilities consider various approaches. The optimum solution will probably be system-specific. If a utility considers GAC adsorption for DBP precursor control, the following approach is recommended based on the findings in this study:

- Detailed and accurate information should be obtained about the pore size distribution and pH_{PZC} values of candidate GACs to be evaluated for DBP precursor control. Carbon manufacturers should be able to provide this information for their products. The conventional parameters such as BET surface area and total pore volume are not useful to evaluate a GAC for DOM removal. Among several candidates GACs, the carbons with large surface areas and volumes in pores >1 nm and basic pH_{PZC} values should be selected for DBP precursor control. Among several GACs with neutral or basic pH_{PZC} values, the GACs having higher pore volumes in pores > 1 nm should be preferred since the effect of surface acidity becomes less important than the pore size distribution above the neutral pH_{PZC}.
- RSSCTs combined with THM and HAA formation tests must be performed to assess and compare the performance of candidate GACs. However, batch constant-dose (isotherm) experiments can be used to evaluate the DOM uptakes of several GACs. The carbons showing the highest DOM uptake at the typical DOC concentration of the water to be treated should be selected for further RSSCTs. Performing these initial batch screening experiments at a constant dose, particle size and contact time for all GACs is important for proper comparison. During the RSSCTs, periodical monitoring of TOC and UV_{254} at the effluent, although useful, is not sufficient because it does not provide information regarding the compliance with the DBP regulations. Therefore, RSSCT effluent must be periodically collected and exposed to post-disinfection at the typical treatment plant and distribution system conditions (i.e., chlorine or chloramines, and typical residence time of the distribution system) to examine THM and HAA formation. The results of RSSCTs can be plotted as a function of operation time for a large-scale column, as done in this study, to determine the operational life of a GAC filter for THM and HAA formation control.
- Removal of high molecular weight components during conventional treatment processes significantly increases the operational time of GAC filters for DBP formation control. Therefore, the impact of conventional treatment processes on GAC adsorption and DBP formation control should be evaluated in designing and operating GAC adsorption systems. All the tests for GAC selection and evaluation of GAC adsorption must be conducted using a typical water (e.g., after conventional treatment processes) that will enter the GAC filter during the practical application. Operating GAC filter at a low pH value (e.g., 5.5 tested in this study) can further increase the DOM removal and DBP formation control. The effects of pH on DOM removal in both conventional treatment processes and subsequent GAC adsorption need to be optimized. In all these investigations, results of THM and HAA formation tests at the simulated treatment plant and distribution system conditions must be taken as the metric to compare the effectiveness of several GACs or GAC adsorption with other alternatives for DBP formation control.

RECOMMENDATIONS FOR FUTURE RESEARCH

There are several research issues that need further investigation:

- The impact of carbon modifications were tested only with one high $SUVA_{254}$ and one low $SUVA_{254}$ water in this study. Since DOM is a source specific heterogeneous mixture, more natural waters from various sources should be tested to further generalize the conclusions from this study.
- The carbon source (i.e., raw material) and its structure should be also considered in the preparation of modified GACs for DOM removal. It is recommended to have a side by side comparison of the GACs produced from various raw materials.
- In order to understand differences in the adsorption patterns of heterogeneous DOM between a batch isotherm and RSSCT, it is necessary to perform further systematic and well-characterized experiments.
- Various carbon surface modification techniques need to be further explored to see if it is possible to further increase the surface affinity of activated carbons toward various DOM components. For example, the effects of basic nitrogen-containing carbon surface functional groups on the DOM removal deserve further investigation. Various nitrogen-containing functionalities can be loaded on the carbon surface by liquid phase treatment. The presence of various positively charged nitrogen functionalities may enhance the DOM removal. In terms of iron impregnation of GACs, it is necessary to understand the affinity of different DOM components to various types of iron species and how to control the formation of such iron species during impregnation of GACs. It should be noted that the dispersion and speciation of iron is more important than the iron content of the impregnated GACs.
- In order to gain insight to the role of DOM characteristics in adsorption by modified activated carbons, different DOM fractionation and characterization methods (e.g., based on hydrophobicity (hydrophobic, hydrophilic and transphilic fraction) or based on size (e.g., ultrafiltration fraction, SEC coupled with UV, TOC and TN detection) can be utilized.
- All future GAC research should include RSSCTs and THM and HAA formation measurements at representative drinking water treatment conditions to assess practical applicability of the findings.

APPENDIX A
PROCESS FLOW DIAGRAMS OF MYRTLE BEACH AND SPARTANBURG WATER TREATMENT PLANTS

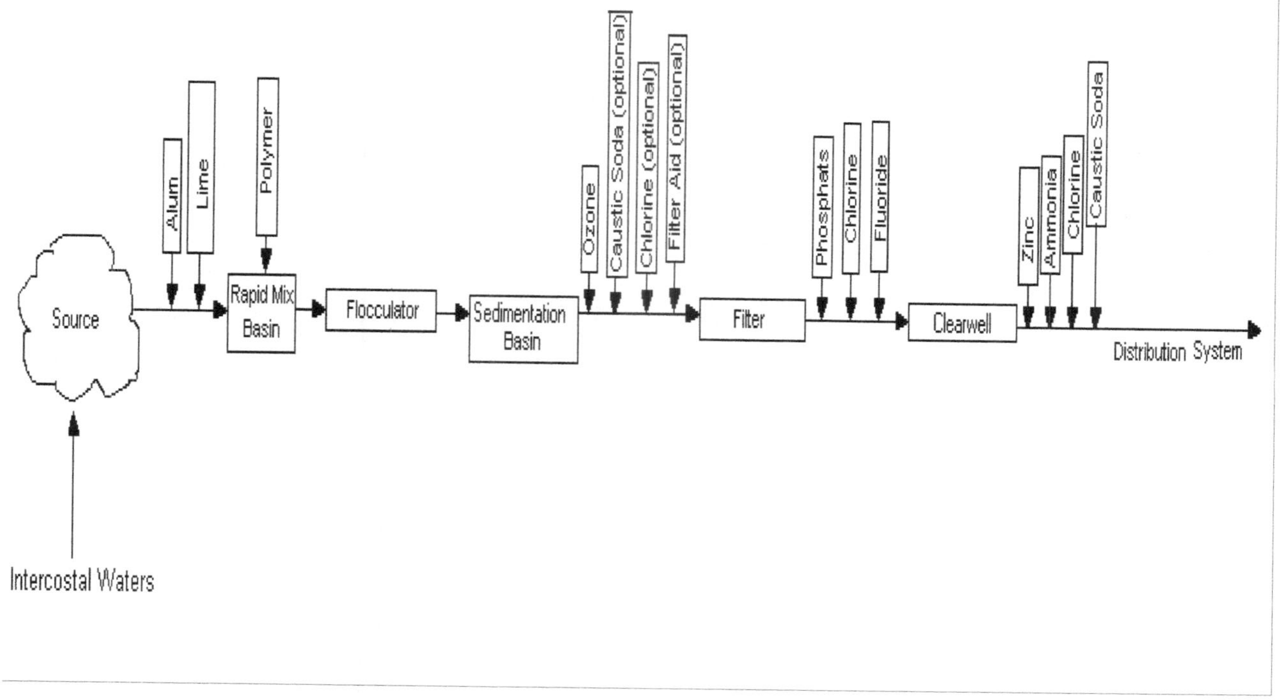

Chemicals	Average dose (mg/L)
Ozone	4.7
Coagulant (Alum)	115.66
Total Lime	24.86
Caustic (Coag)	5.58
Caustic	19.34
Polymer (Coag)	0.17
Ammonia	0.81
Free Cl_2	1.53
Total Cl_2	3.23
Fluoride	0.82
Carbon	0.09
Inhibitor (Pyro)	1.23
Inhibitor (Zinc)	0.20

Figure A.1 Process flow diagram for Myrtle Beach drinking water treatment plant.

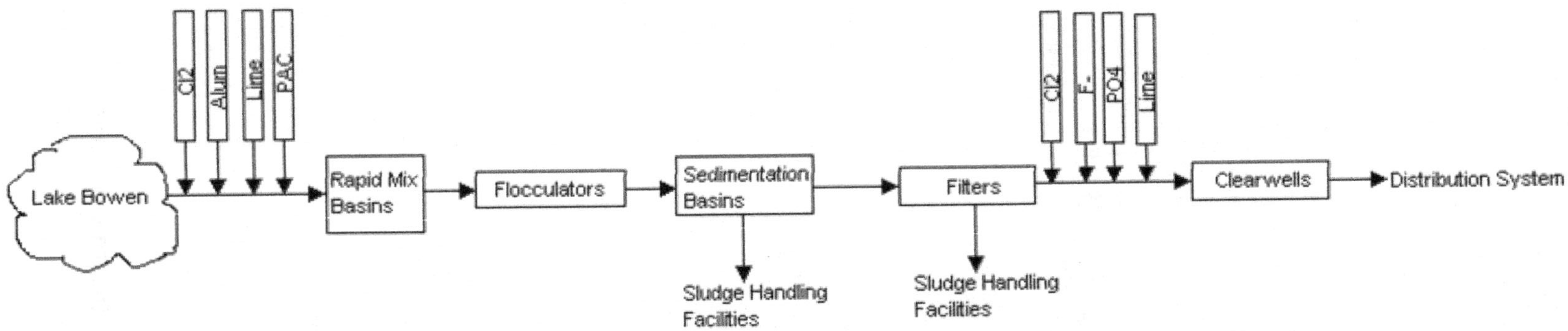

Chemicals	Typical dose (mg/L)
Pre-chlorine	2.5 – 4.5
Coagulant (Alum)	15 - 30
Pre - Lime	2 - 15
Powdered Activated Carbon (PAC)	2 - 15
Polymer (Coagulation)	0.2 – 0.5
Post-chlorine	0.75 – 1.5
Fluoride	0.8 – 1.2
Inhibitor (zinc hexametaphosphate blend)	0.8 – 1.2
Post-lime	5 - 15

Figure A.2 Process flow diagram for Spartanburg drinking water treatment plant.

APPENDIX B
DESIGN OF RAPID SMALL SCALE COLUMN TESTS

The design of RSSCT used the guidelines in the manual published by USEPA for the ICR studies (USEPA, 1996). Since detailed information is available in this manual, it will not be repeated again here. In this section, only a brief outline of the design will be presented along with some characterization work performed for the RSSCT set-ups assembled for the project. The variables used in the equations and calculations are defined at the end of this section.

BACKGROUND

This design is based on the proportional diffusivity (PD) design approach. The following relationships are used between the parameters of large scale column and small scale column.

$$SF = \frac{d_{LC}}{d_{SC}}$$

$$EBCT_{SC} = \frac{EBCT_{LC}}{SF}$$

$$V_{SC} = \frac{Re_{SC,\,\min} \times v_{SC} \times \varepsilon_{SC}}{d_{SC}}$$

Requirements for particle size and hydraulic loading (superficial velocity)

a. To avoid channeling, the ratio of column inner diameter to particle size must be greater than 25.
b. To assure that intraparticle diffusion controls the mass transfer of DOM, the Reynolds number (Re) is recommended to range between 0.5 and 1.0.

Design requisitions

$EBCT_{LC}$ = 10 min
SF = 10
d_{sc} = 0.150mm (100 mesh)< d_{sc} <0.106mm (140 mesh)
C_0 = 3 mg TOC/L
ε = 0.4 (assumed)
$\nu_{sc} = 1.0 \times 10^{-6}$ m^2/s (at 20°C)
ρ_b= 0.5 g/cm^3 (assumed)

The determination of column parameters

1) Inner diameter of column: D_{SC}

$D_{sc} > 25d_{sc} = 25 \times 0.149 = 3.73$ mm
Column with inner diameter of 7 mm was selected because of its availability in our laboratory.

2) Hydraulic loading (superficial velocity): v_{sc}

$v_{sc} = Re_{sc,min} \times \nu_{sc} \times \varepsilon_{sc} / d_{sc}$
$= (0.5 \sim 1) \times 16.1$ cm/min (for d_{sc}=0.149mm)
The value of 15 cm/min was selected for v_{sc}.

3) The length of carbon bed:

$EBCT_{SC} = EBCT_{LC} \div SF = 1$ min.
$l_{SC} = EBCT_{SC} \times v_{sc} = 15$ cm

4) Carbon usage:

$m_{sc}=\rho_b \times \pi/4 \times D_{sc}^2 \times lsc$

5) Flow rate:

$Q_{sc} = v_{sc} \times 3.14 \times D_{sc}^2 / 4 = 8.31$ L/day

6) The operation time:

- Empirical formula for TOC (DOM)

$t_{sc} = 2 \times 21700 C_0^{-1.3} EBCTsc=7.2$ days

- Estimated from equilibrium adsorption capacity assuming instantaneous mass transfer

$t_{sc}^t= m_{sc} \times q_e/(C_0 \times Q_{sc}) = 3.49$ days
If safety factor of 4 is considered:
$t_{sc}^t=14$ days

- Previous RSSCT results show that usually it takes about two weeks to get the breakthrough curves for other DOM.

7) Required influent volume:

$V_{sc}= t_{sc}^t \times Q_{sc} = 14 \times 8.31=116$ L

8) DOM consumption per test

$m_{DOM} =C_0 \times V_{sc}=349$ mg as TOC

Design Summary of RRSCT Characteristics

$EBCT_{LC}$:	10 min
Scaling Factor:	10
$EBCT_{SC}$:	1 min
l_{SC}:	15 cm
v_{SC}:	15 cm/min
D_{SC}:	7 mm
Q_{SC}:	5.8 mL/min
$Re_{SC,min}$:	1.0
d_{sc}:	0.150mm (100 mesh)< d_{sc} <0.106mm (140 mesh)

Dispersion Tests

To study the flow patterns in RSSCTs and the effect of column packing, two tracer tests were conducted under the conditions of the DOM experiments but using DDW and sodium chloride (NaCl), as the tracer. In these tests, two independent RSSCTs (i.e., two independently packed columns and two separate pump set-ups) were used. The cumulative residence time distribution (*F*-curve) for the two independent RSSCTs was obtained using conductivity detection of NaCl (Figure B.1). The results showed that different experimental set-ups (including packed columns and pumps) for the experiments produced very similar flow patterns. Applying numerical differentiation to experimental results of F curve, the exit age or residence time distribution curve (i.e., E-curve, Figure B.2) was obtained. It is evident from the E-curve that channeling was not an issue in the RSSCTs.

A plug-flow with dispersion model was used to simulate the tracer tests results. Mean residence time (t_m) and variance (σ^2) were obtained from the tracer tests by numerical integration method. The dispersion coefficient (D) and the Peclet number (Pe) for the two columns were calculated. Following are the corresponding equations used in these calculations:

$$E(t)=dF(t)/dt$$

$$t_m=\int t dF(t)$$

$$\sigma^2=\int t^2 dF(t)-t_m^2$$

$$\sigma^2/t_m^2=2(D/uL)-2(D/uL)^2\{1-e^{-(uL/D)}\}$$

(closed vessel entrance and exit conditions)

$$Pe=uL/D$$

The Peclet number is the product of system length and fluid velocity, divided by diffusivity. It is often used instead of the dispersion number. If Pe >>1, the mass transfer caused by dispersion is small compared to advection. If Pe<<1, dispersion dominates the mass transport process. For the RSSCTs, the calculated Peclet numbers were 29 and 32 for the two independent tests, and the mean residence time was approximately 6.7 minutes. These results show that dispersion was not significant in the RSSCT experimental set-ups designed and prepared for the study. The calculated Peclet numbers and residence times were used in the plug-flow dispersion to simulate the F-curve and compare to the experimental results (Figure B.3).

From this comparison, it is evident that the calculated cumulative residence time distribution can simulate corresponding experimental results very well, except at short elution times. This is probably due to limited number of samples collected at the beginning of the experiments with low conductivity values.

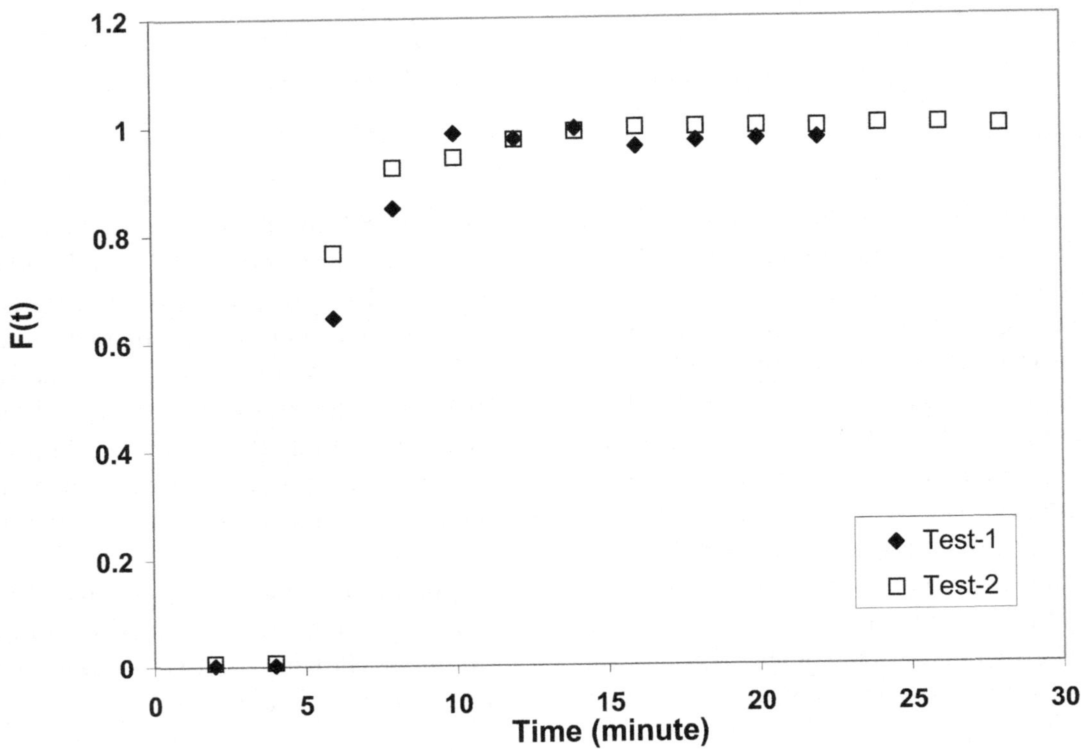

Figure B.1 Step-input tracer response F curve of RSSCTs

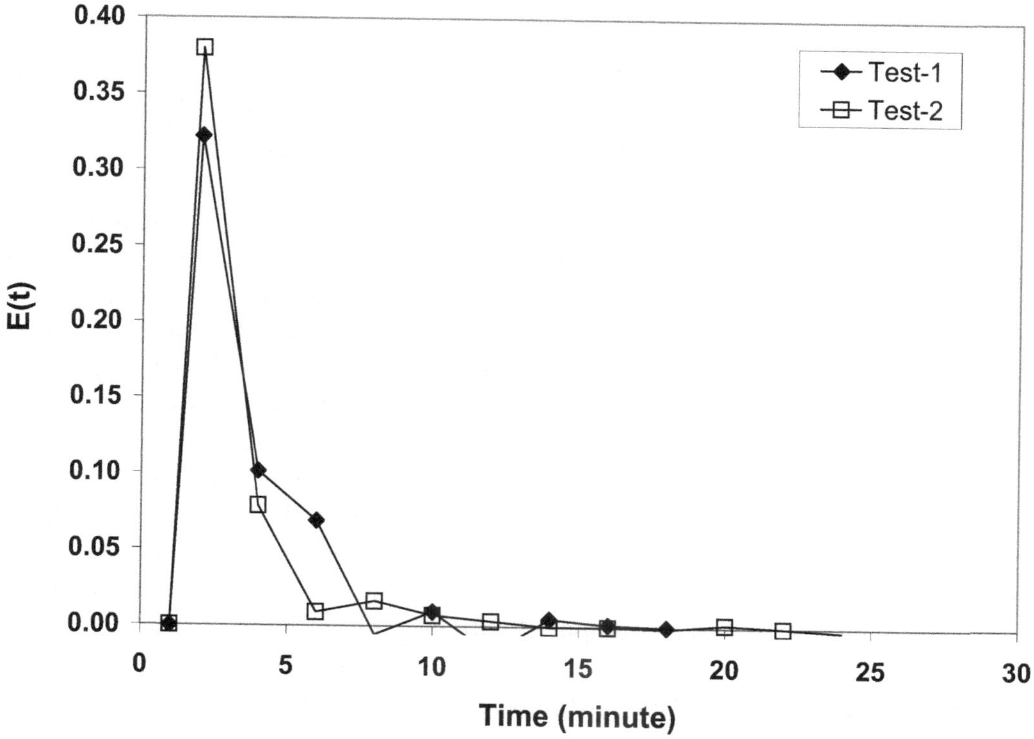

Figure B.2 Residence time distribution curves (E-curves) of the RSSCTs

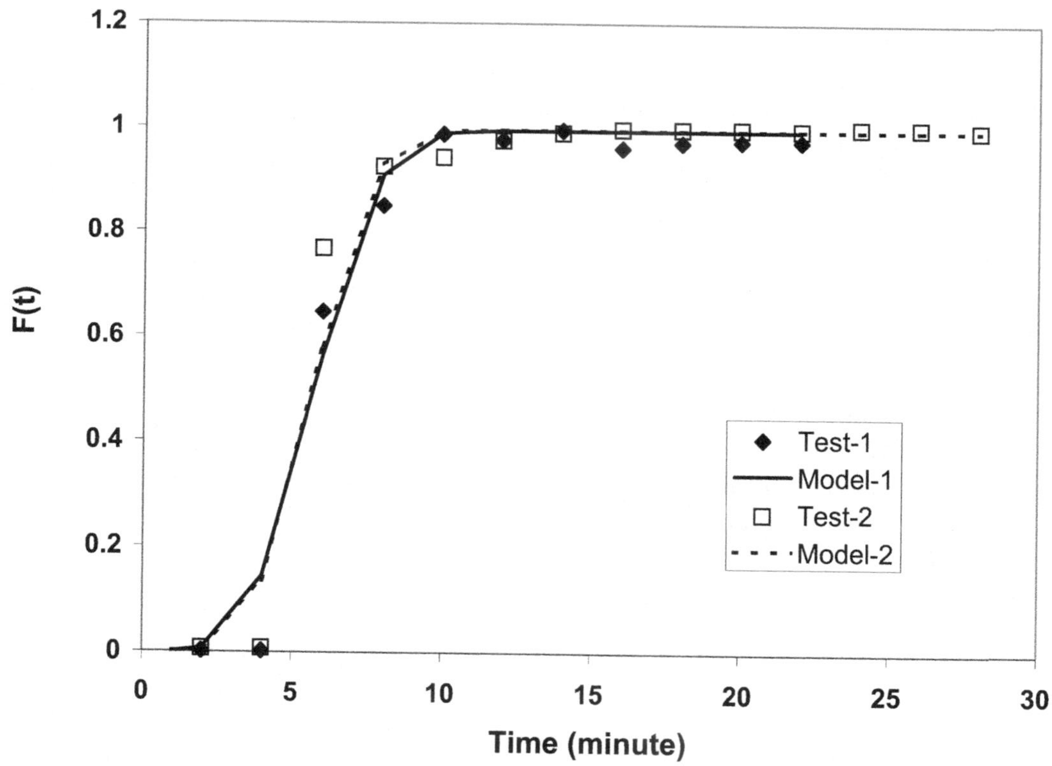

Figure B.3 Simulation of experimental results using a Plug-Flow dispersion model

Reproducibility Tests

The tracer test results showed that the techniques developed in this study will provide reproducible flow conditions for the RSSCT set-ups to be used in this study. However, there are a relatively large number of non-quantifiable factors that had the potential for impacting the DOM breakthrough patterns. This includes variability in sampling glassware cleanness, variability in the analytical measurements from day-to-day operation, the fluctuations in flow rate and pressure change during long experiment periods, and differences in column packing. Therefore, to evaluate the reproducibility of breakthrough patterns, two RSSCTs were performed using the virgin HD and Myrtle Beach Raw water DOM side-by-side. Both RSSCTs were fed with the same stock solution of 3.2 mg TOC/L during the experiments. The UV_{254} and TOC breakthrough results are shown in Figure B.4. The results showed a good reproducibility. Logarithmic equations were fitted to the experimental data and the variability between the two breakthrough curves (i.e., Test-1 vs Test-2) were calculated at the same bed volumes. The variability as UV_{254} or TOC was within 5%. These results along with the previously discussed tracer study results indicated that using the experimental set-ups designed and built for this project, it is possible to obtain reproducible results from RSSCTs. Therefore, results obtained from different column experiments can be compared with each other.

Legend for the variable used in the RSSCT design and calculations

RSSCT: Rapid small scale column test
SF: Scaling factor (the ratio of the carbon particle size used in large scale experiment and RSSCT)
EBCT: Empty bed contact time
$EBCT_{LC}$: Empty bed contact time of large column simulated by RSSCT
$EBCT_{SC}$: Empty bed contact time of small column (used in RSSCT)
d: the diameter of carbon particle
Subscript LC: large column simulated by RSSCT
Subscript SC: small column used in RSSCT
v_{sc}: hydraulic loading (superficial velocity)
Re: Reynolds number

$$\mathrm{Re} = \frac{d * V * \varepsilon}{\nu}$$

ε: the porosity of carbon bed
ν: kinematic viscosity of water
$Re_{sc,min}$: the minimum Reynolds number required by the RSSCT so that internal diffusion is the controlling step of mass transfer process
L: the length of the carbon bed
u: the superficial velocity.
m: mass of carbon used for packing one column
ρ_b: carbon bed density
Q: flow rate of influent
t_{sc}: total operation time of RSSCT

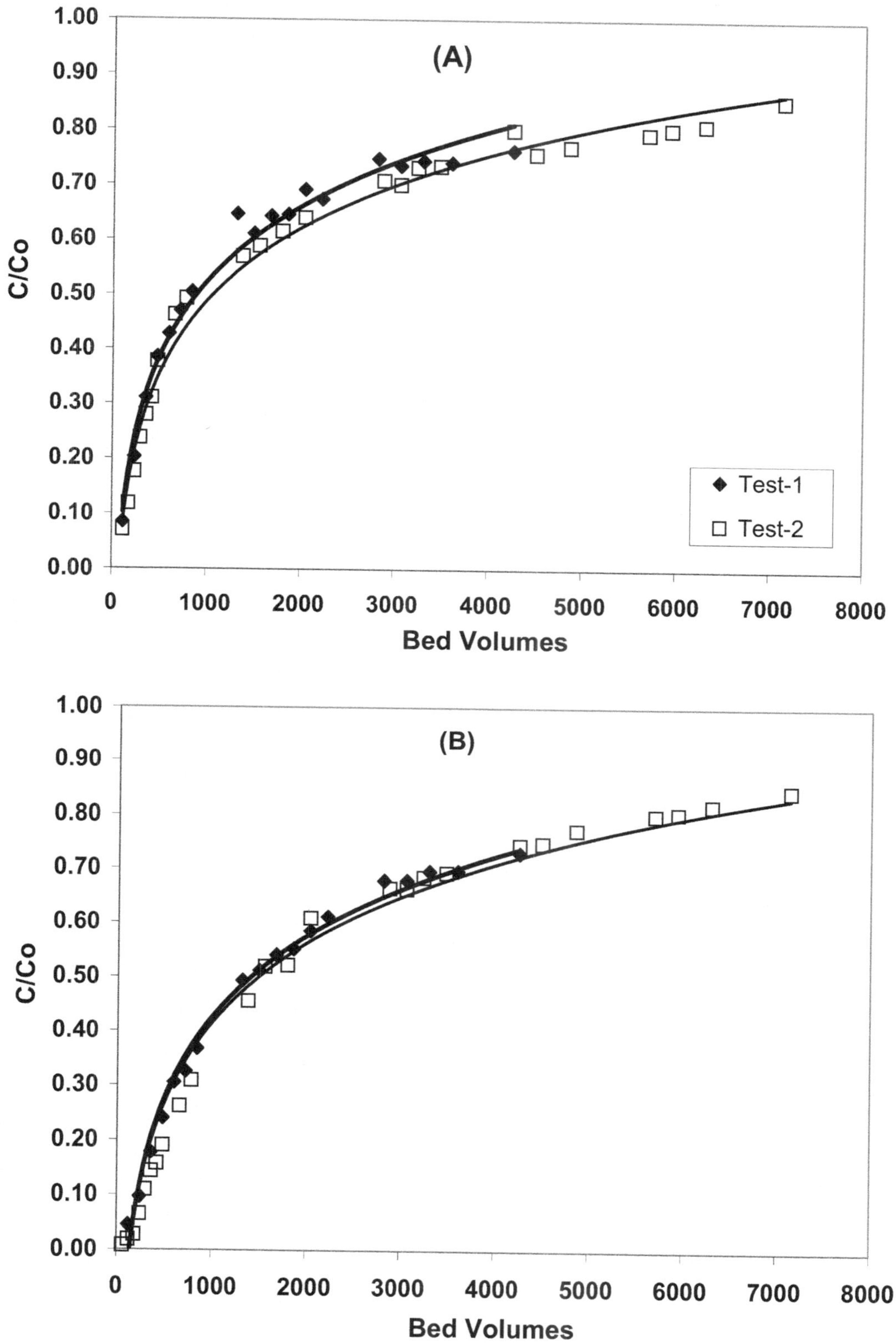

Figure B.4 Column reproducibility test results with virgin HD and Myrtle Beach raw water DOM as (A) UV_{254} and (B) TOC. Lines show logarithmic fits to the data for comparison purpose.

APPENDIX C
STEAM-TREATED GAC RESULTS

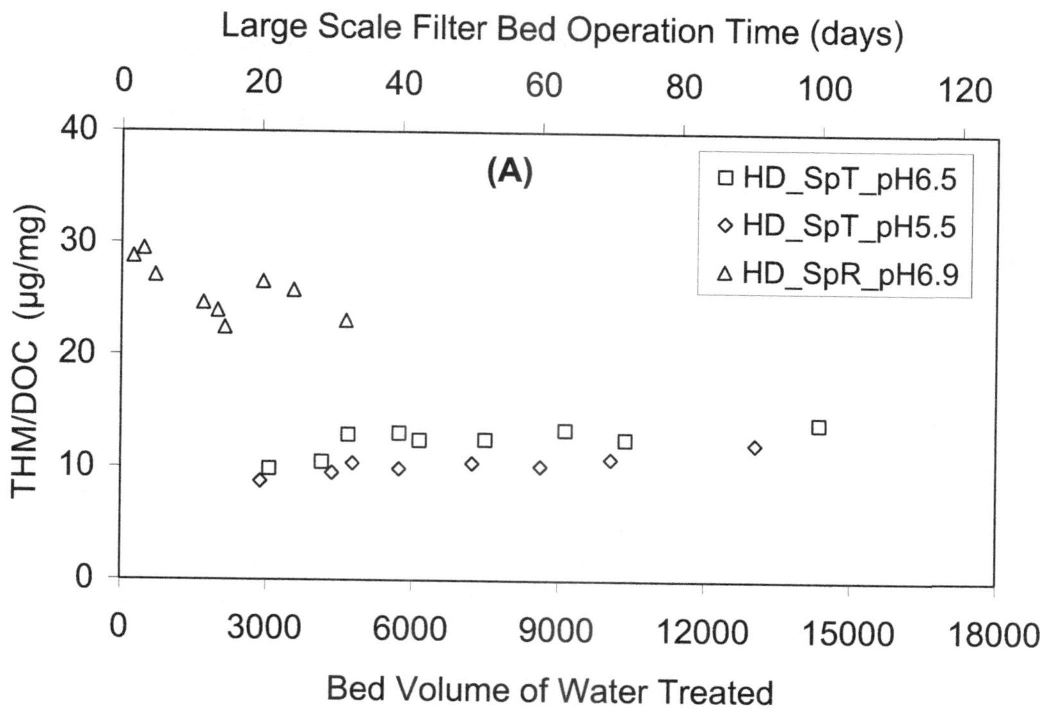

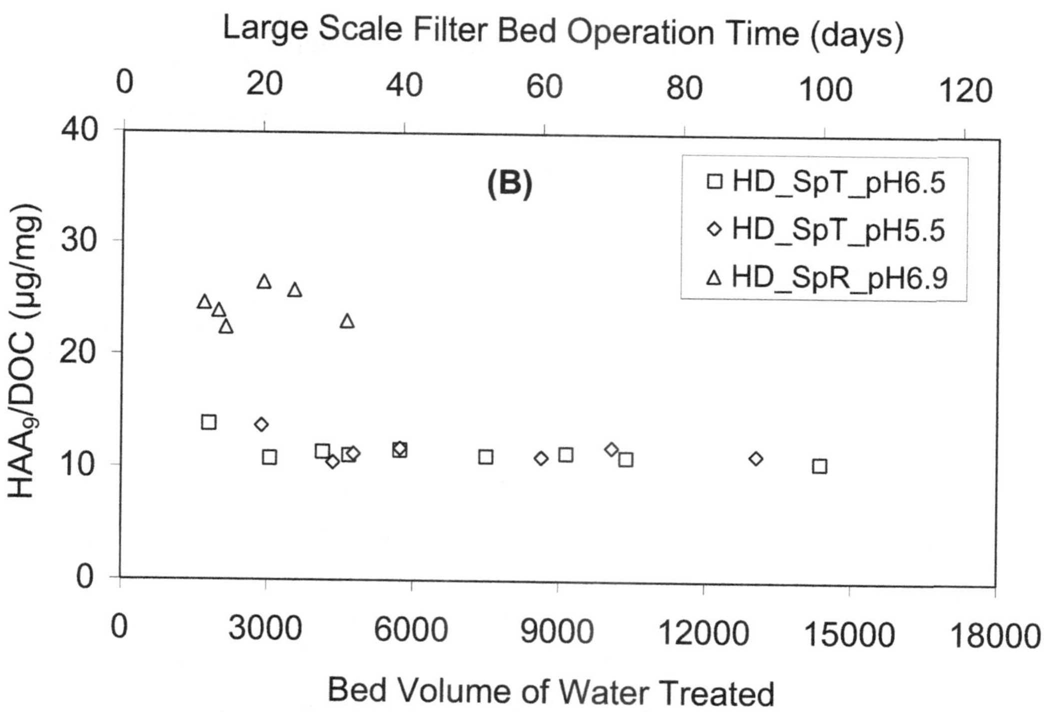

Figure C.1 (A) THM and (B) HAA_9 reactivity of Spartanburg water DOM eluting from the virgin carbon columns

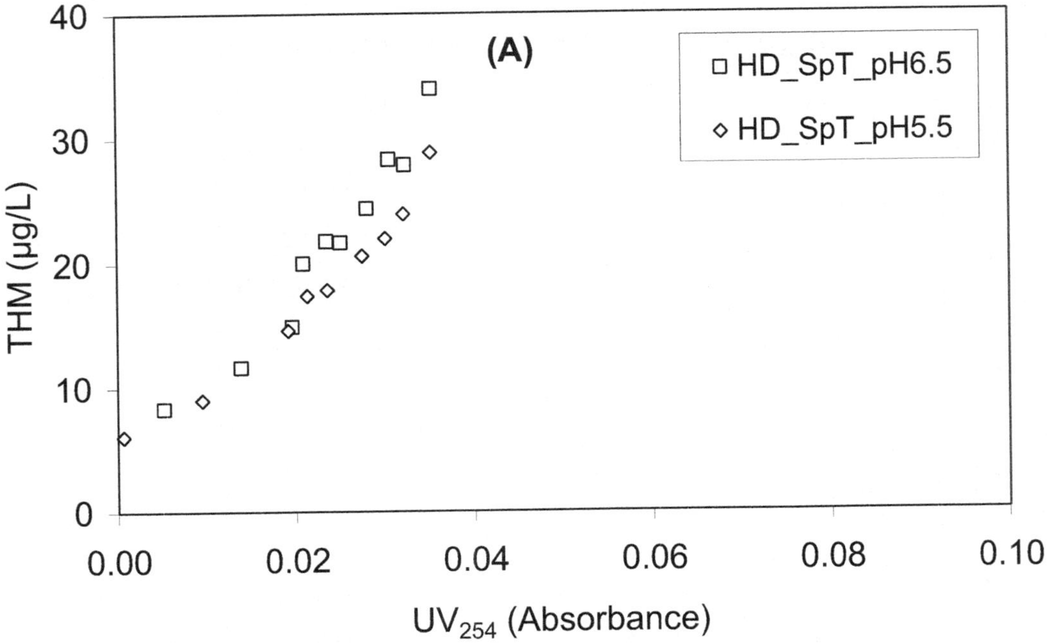

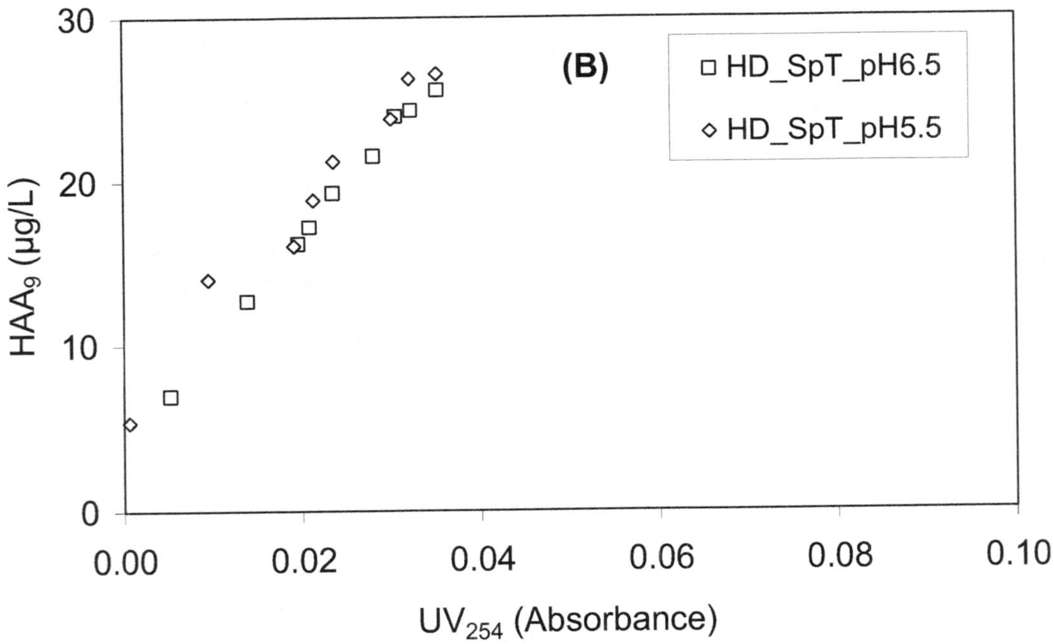

Figure C.2 THM vs. UV_{254} and HAA_9 vs. UV_{254} of Spartanburg water DOM eluting from the virgin carbon columns

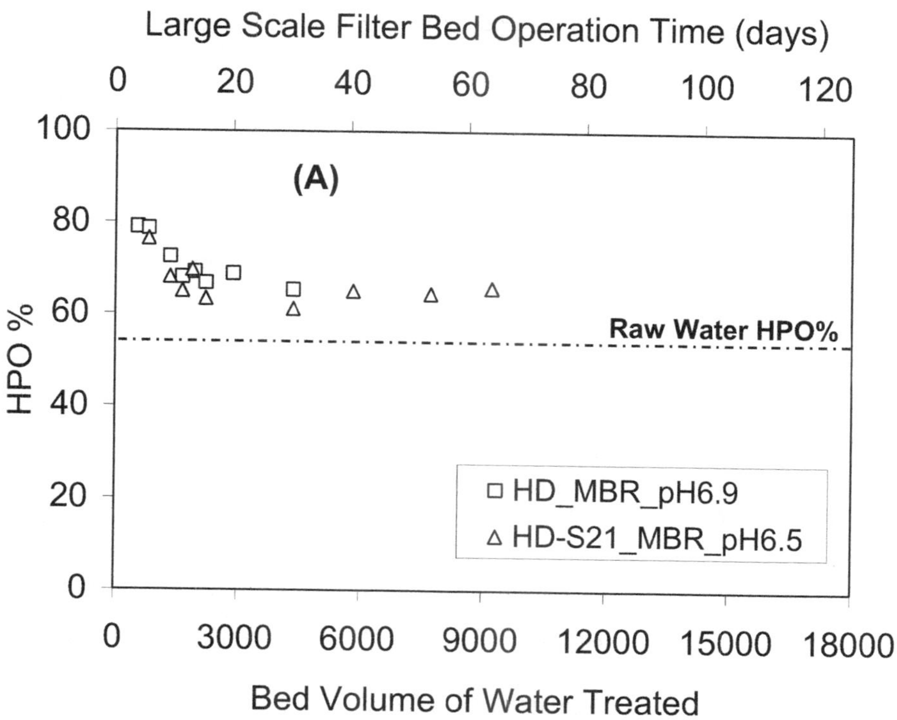

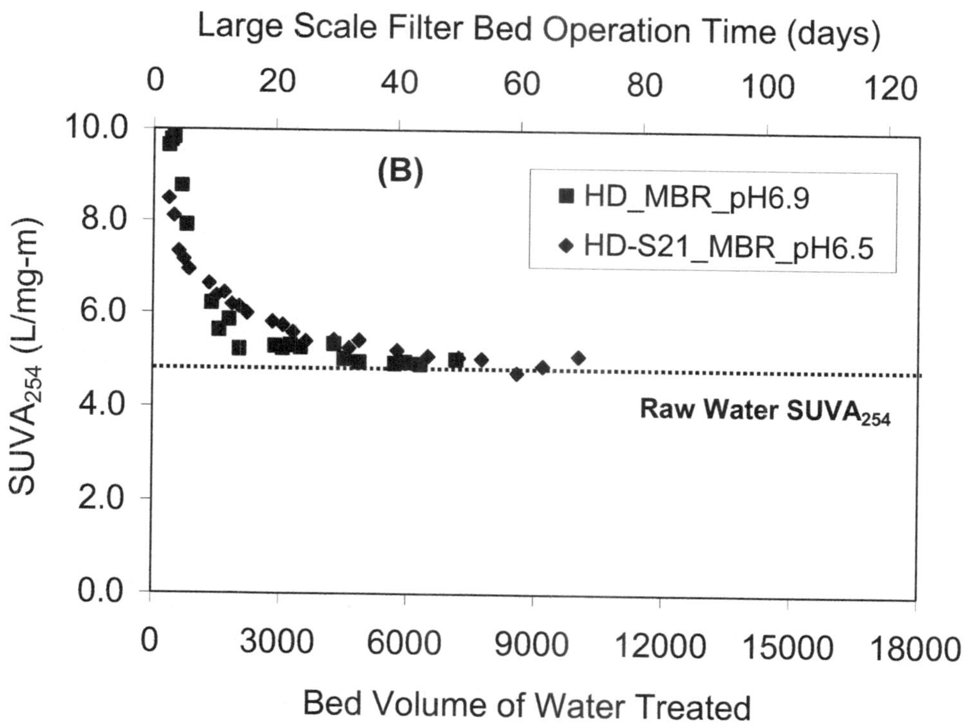

Figure C.3 Myrtle Beach raw water DOM breakthrough profiles, (A) HPO% and (B) $SUVA_{254}$, for virgin and steam-treated carbons

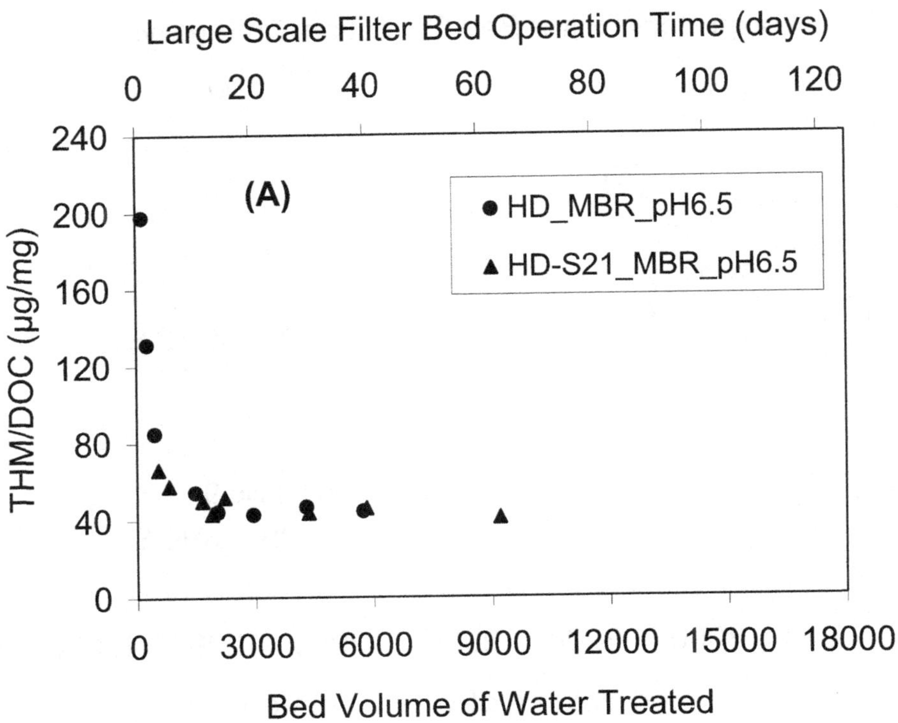

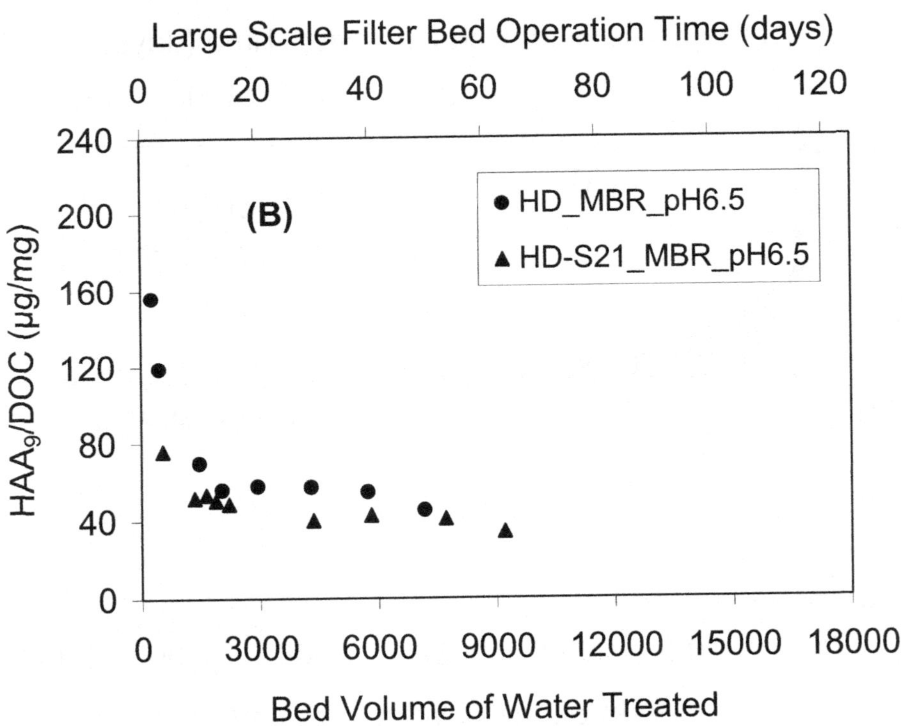

Figure C.4 (A) THM and (B) HAA_9 reactivity of Myrtle Beach raw water DOM eluting from the virgin and steam-treated carbon columns at pH 6.5

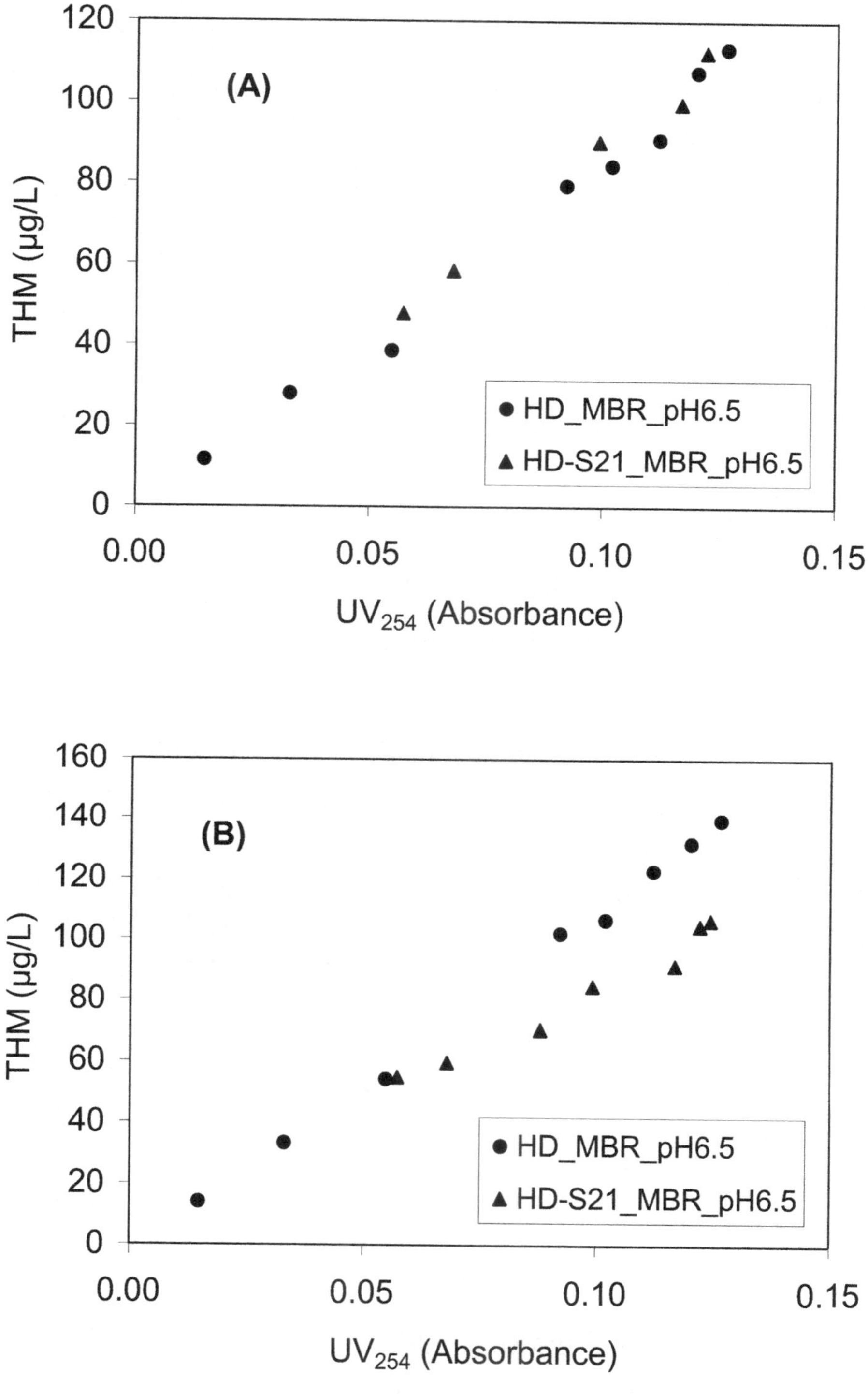

Figure C.5 (A) THM vs. UV_{254} and (B) HAA_9 vs. UV_{254} of Myrtle Beach raw water DOM eluting from the virgin and steam-treated carbon columns at pH 6.5

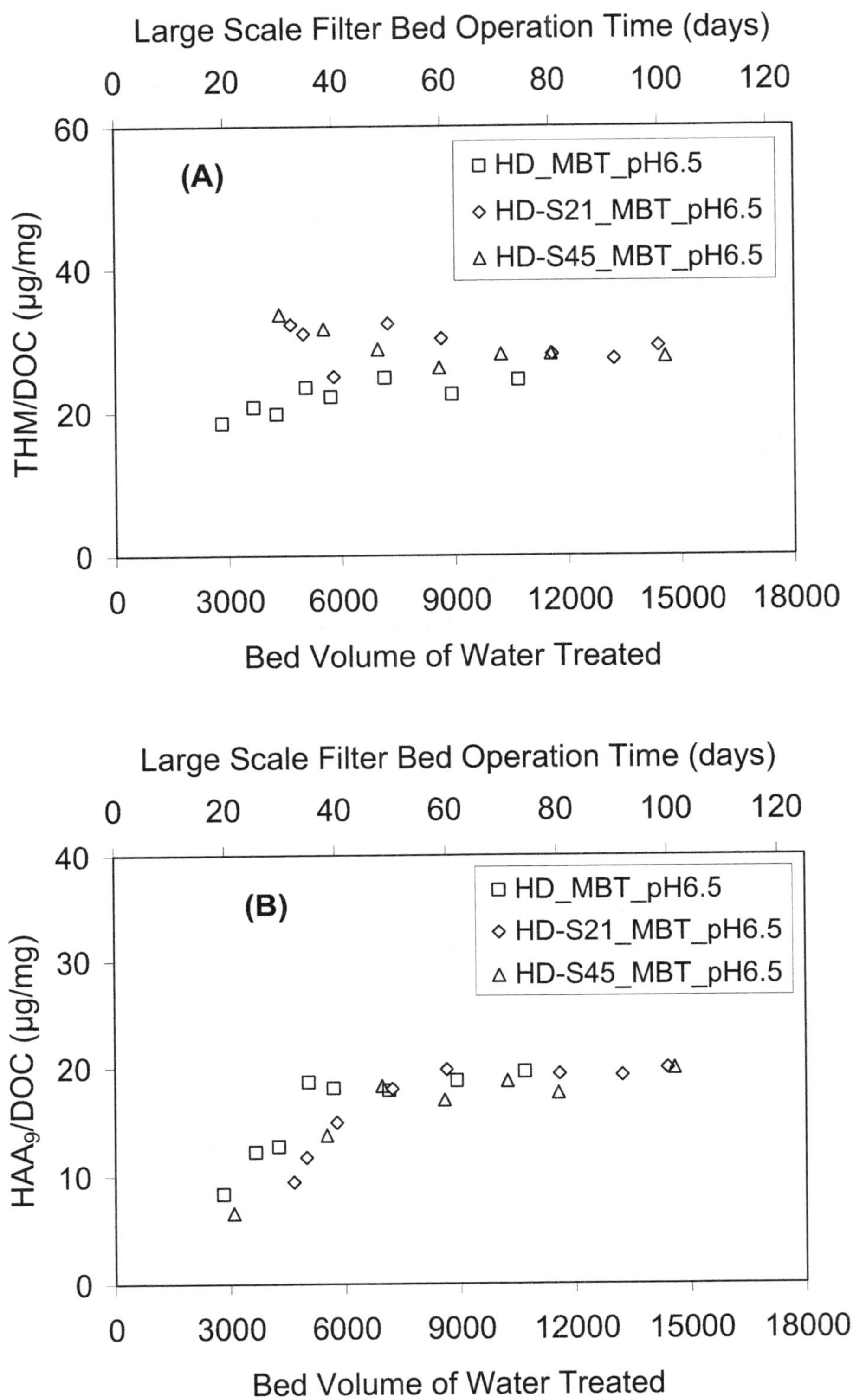

Figure C.6 (A) THM and (B) HAA_9 reactivity of Myrtle Beach treated water DOM eluting from the virgin and steam-treated carbons

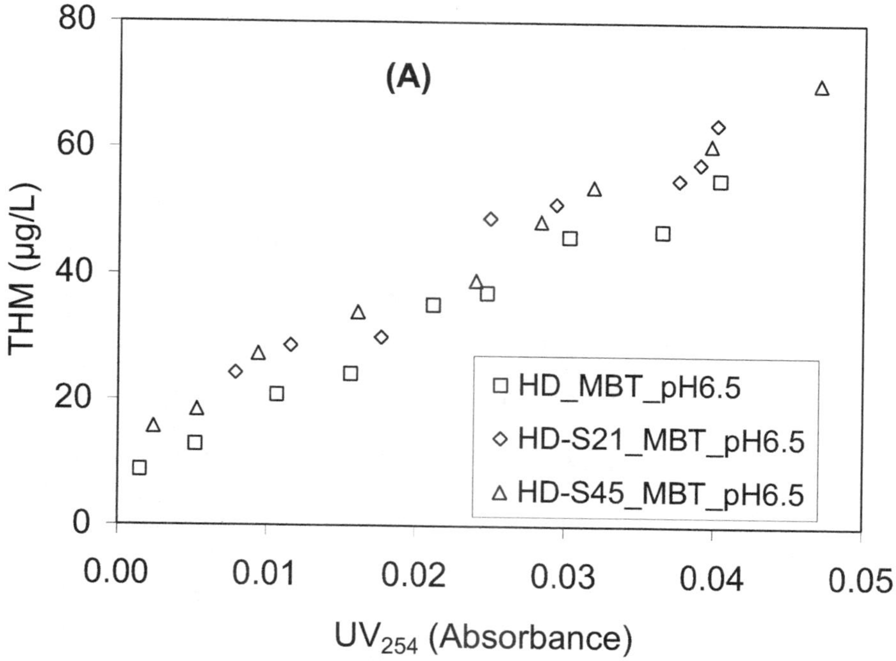

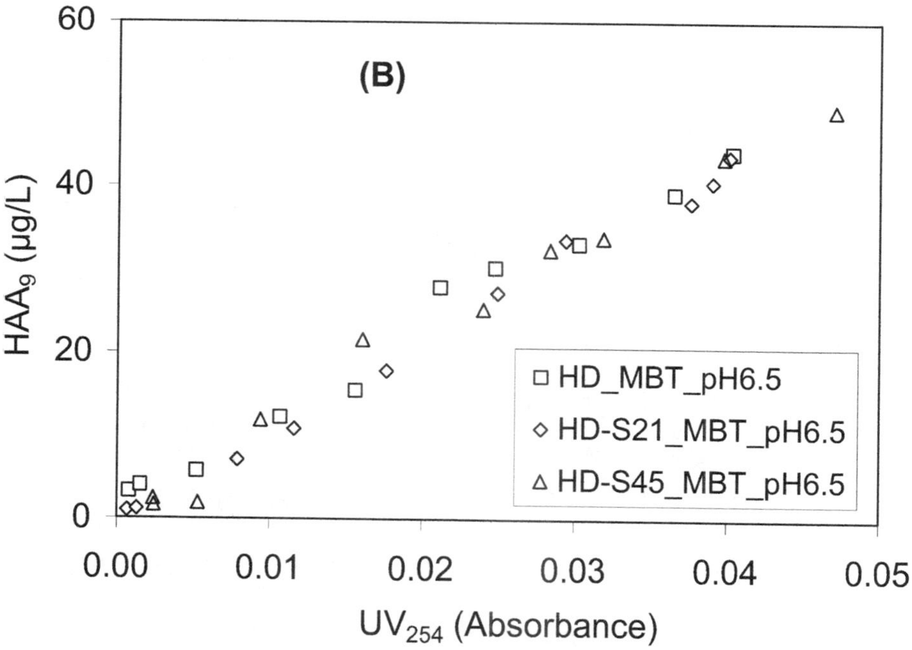

Figure C.7 (A) THM vs. UV254, and (B) HAA_9 vs. UV_{254} of column effluent samples for steam-treated carbons on Myrtle Beach treated water DOM

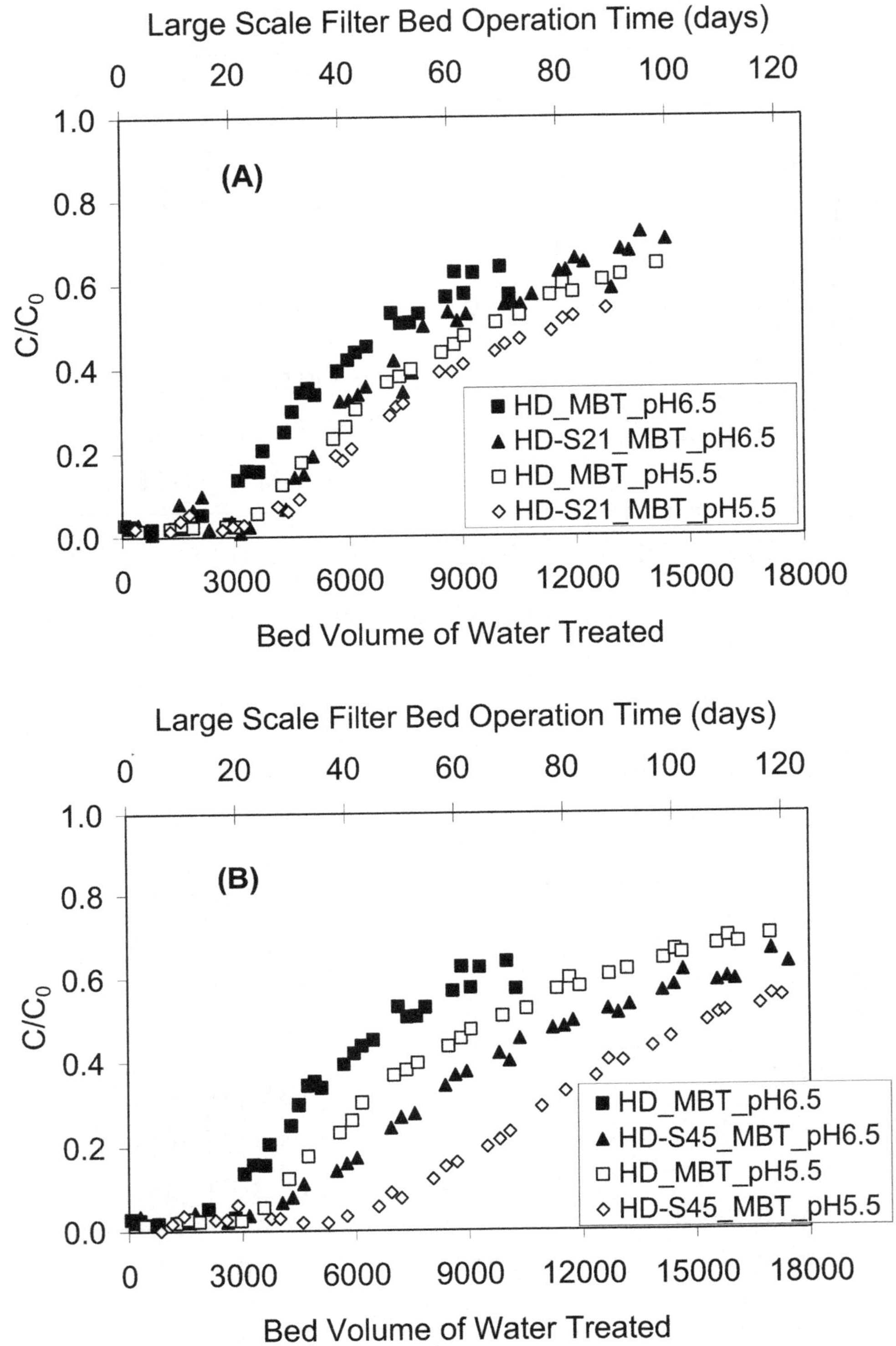

Figure C.8 Myrtle Beach treated water DOM UV_{254} breakthrough profiles for (A) HD-S21 and (B) HD-S45 carbons at pH 6.5 and 5.5

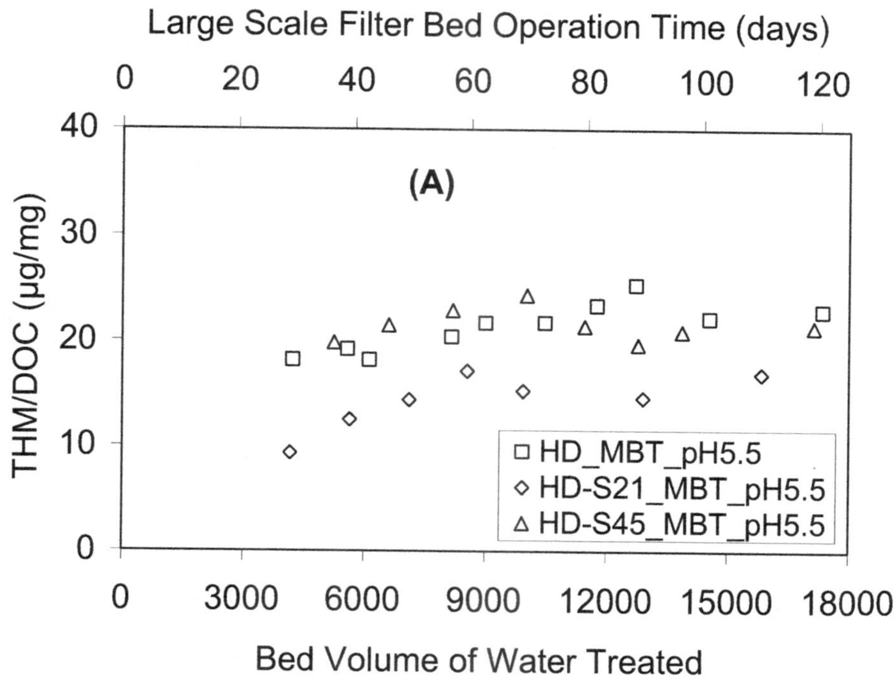

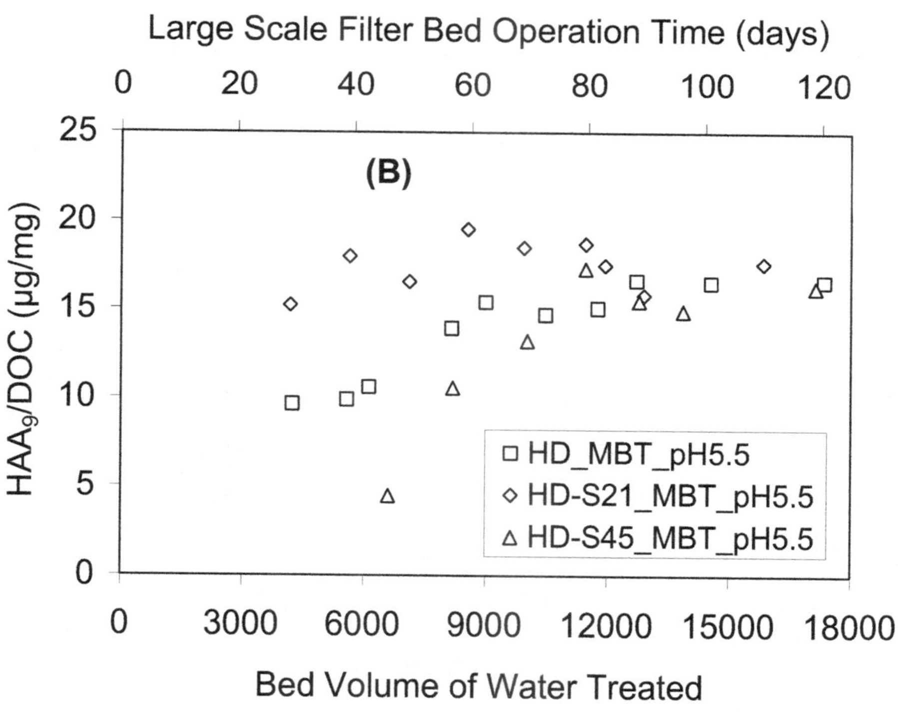

Figure C.9 (A) THM and (B) HAA_9 reactivity of Myrtle Beach treated water from the virgin and steam-treated carbon columns at pH 5.5

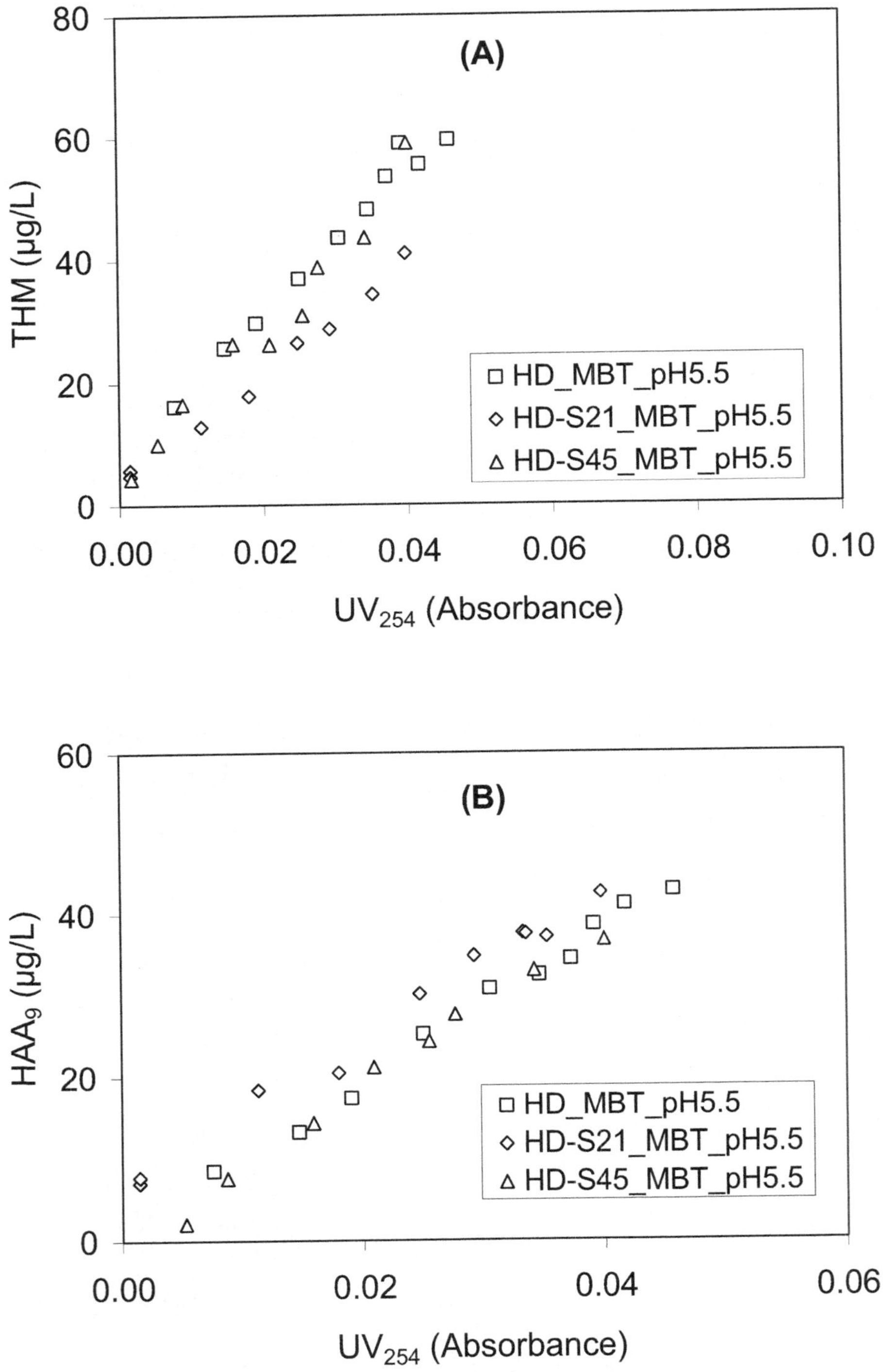

Figure C.10 (A) THM vs. UV_{254} and (B) HAA_9 vs. UV_{254} of Myrtle Beach treated water from the virgin and steam-treated carbon columns at pH 5.5

APPENDIX D
IRON-IMPREGNATED GAC RESULTS

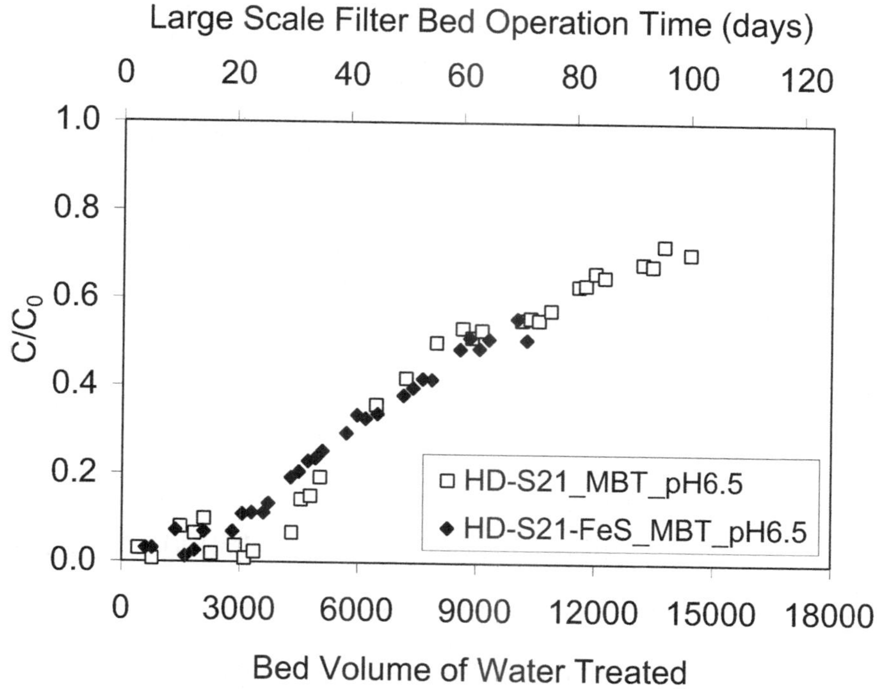

Figure D.1 Myrtle Beach treated water UV_{254} breakthrough profiles for HD-S21 and HD-S21-FeS

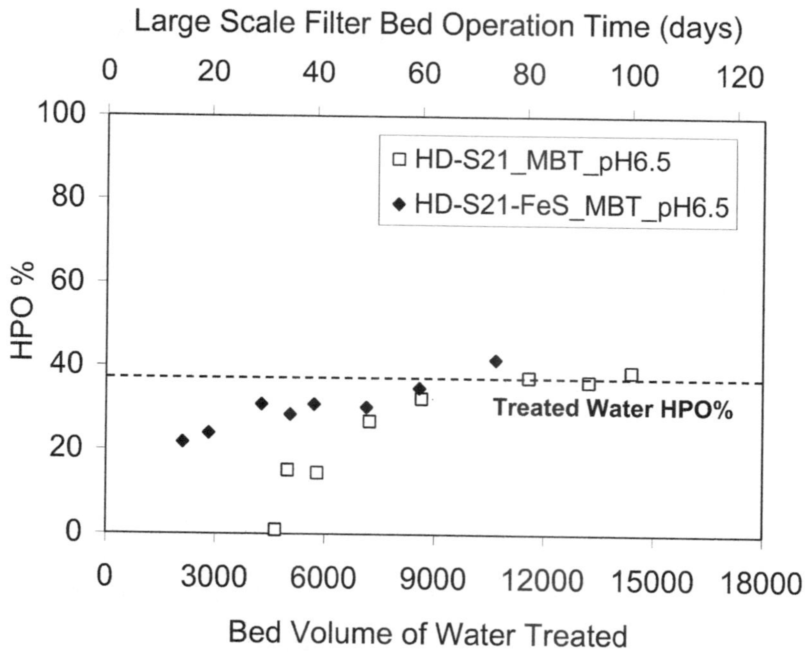

Figure D.2 Myrtle Beach treated water HPO% breakthrough profiles for HD-S21 and HD-S21-FeS

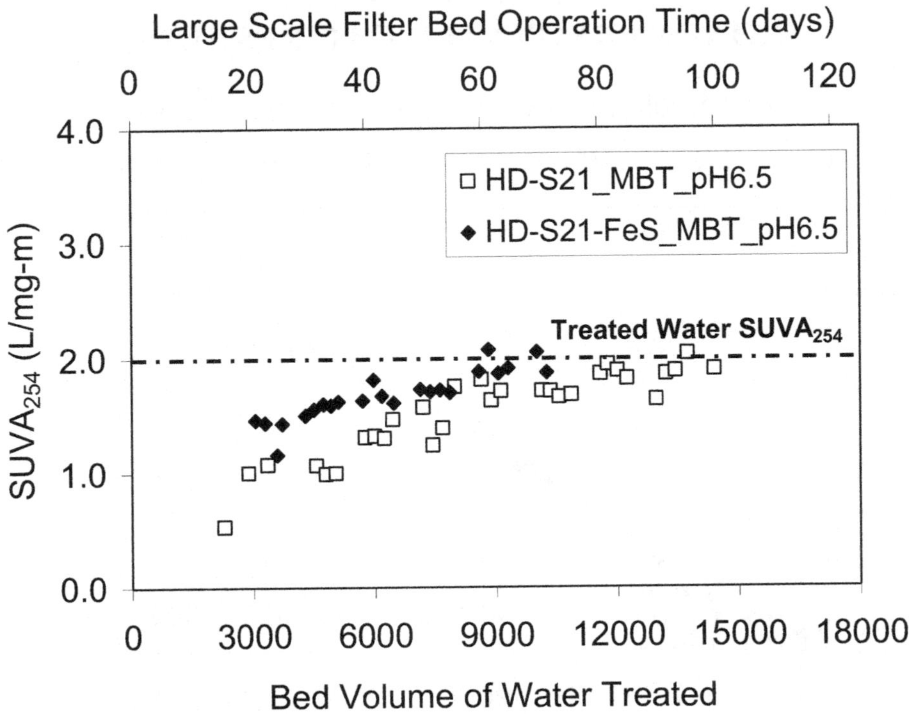

Figure D.3 Myrtle Beach treated water $SUVA_{254}$ breakthrough profiles for HD-S21 and HD-S21-FeS

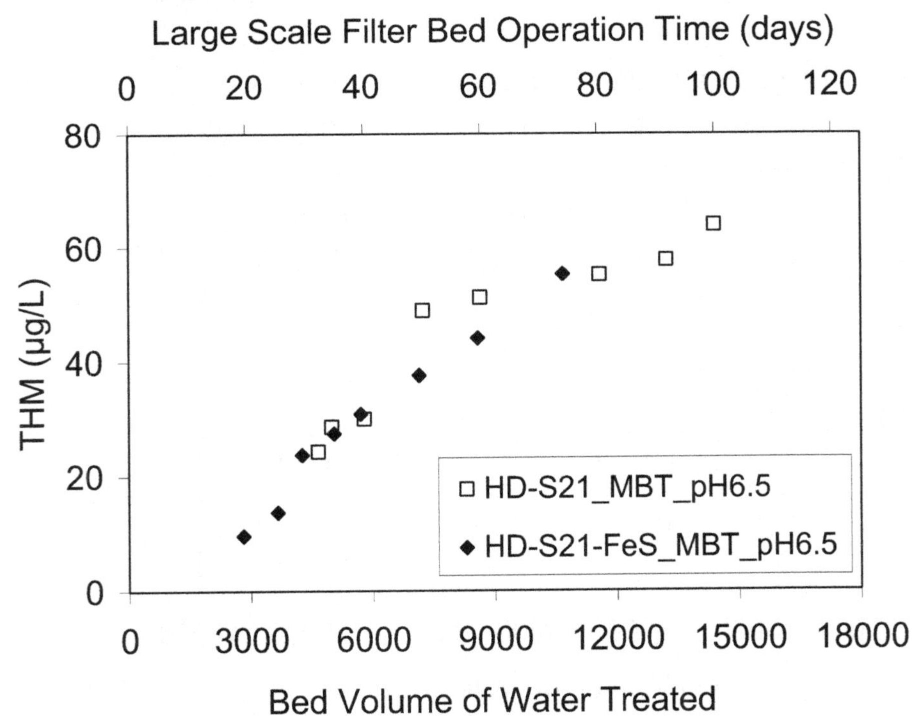

Figure D.4 Myrtle Beach treated water DOM THM formation breakthrough profiles for HD-S21 and HD-S21-FeS

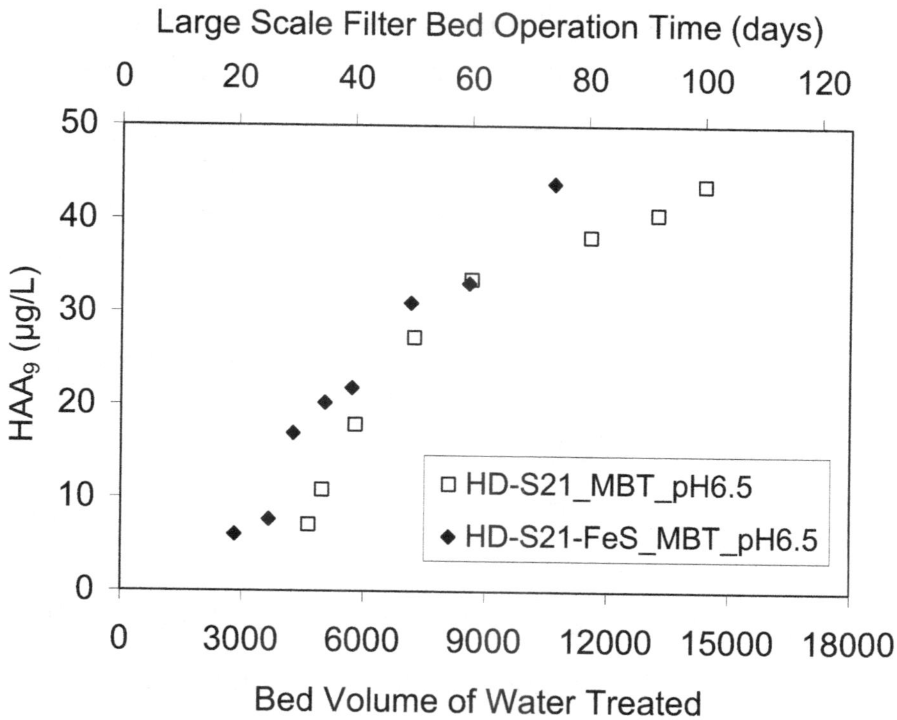

Figure D.5 Myrtle Beach treated water HAA_9 formation breakthrough profiles for HD-S21 and HD-S21-FeS

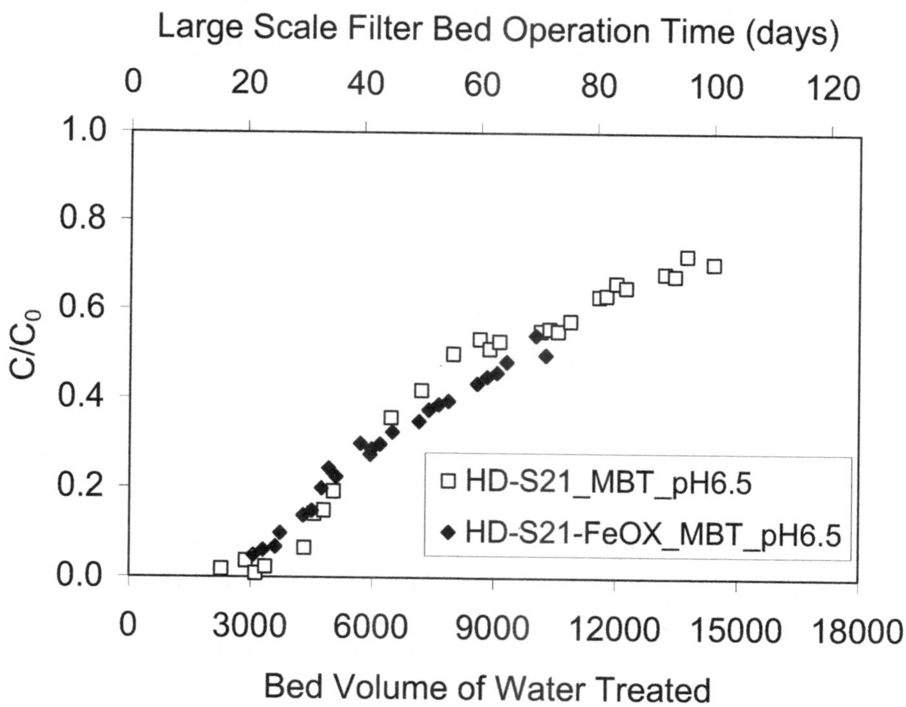

Figure D.6 Myrtle Beach treated water UV_{254} breakthrough profiles for HD-S21 and HD-S21-FeOX

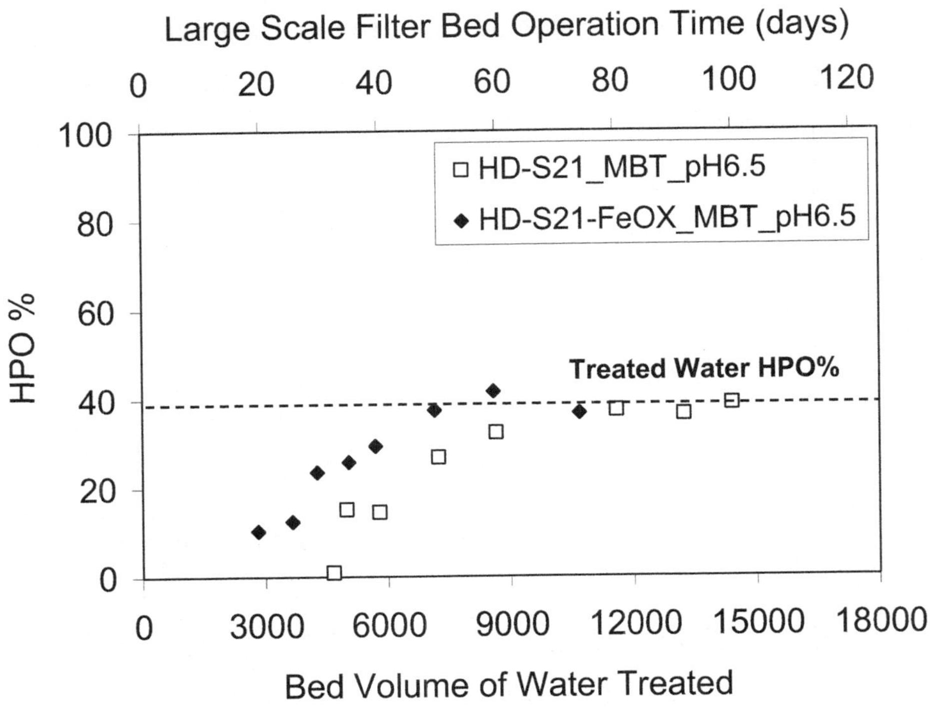

Figure D.7 Myrtle Beach treated water HPO% breakthrough profiles for HD-S21 and HD-S21-FeOX

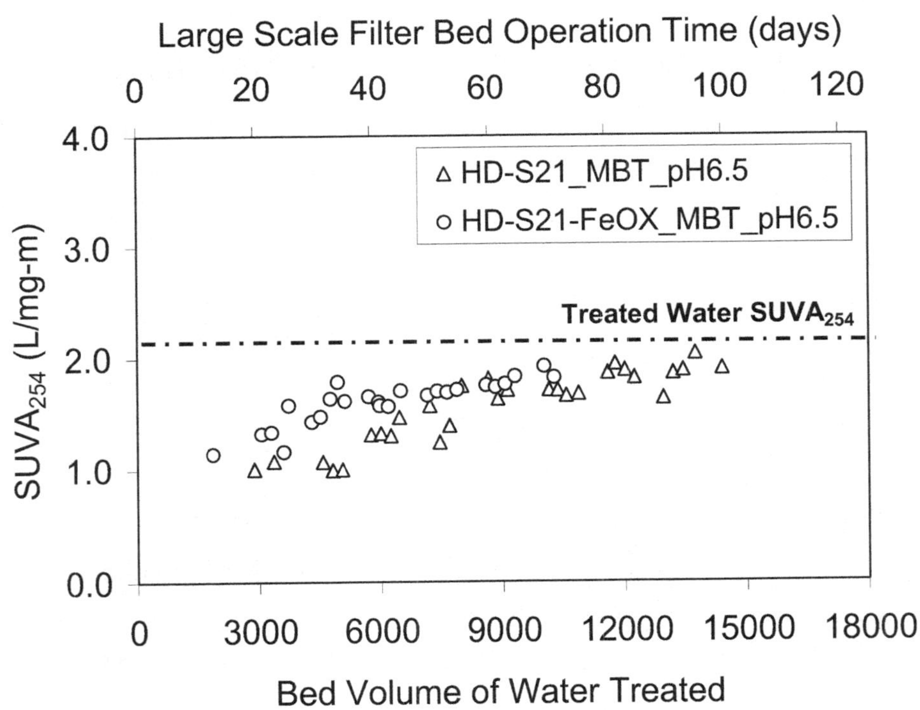

Figure D.8 Myrtle Beach treated water $SUVA_{254}$ breakthrough profiles for HD-S21 and HD-S21-FeOX

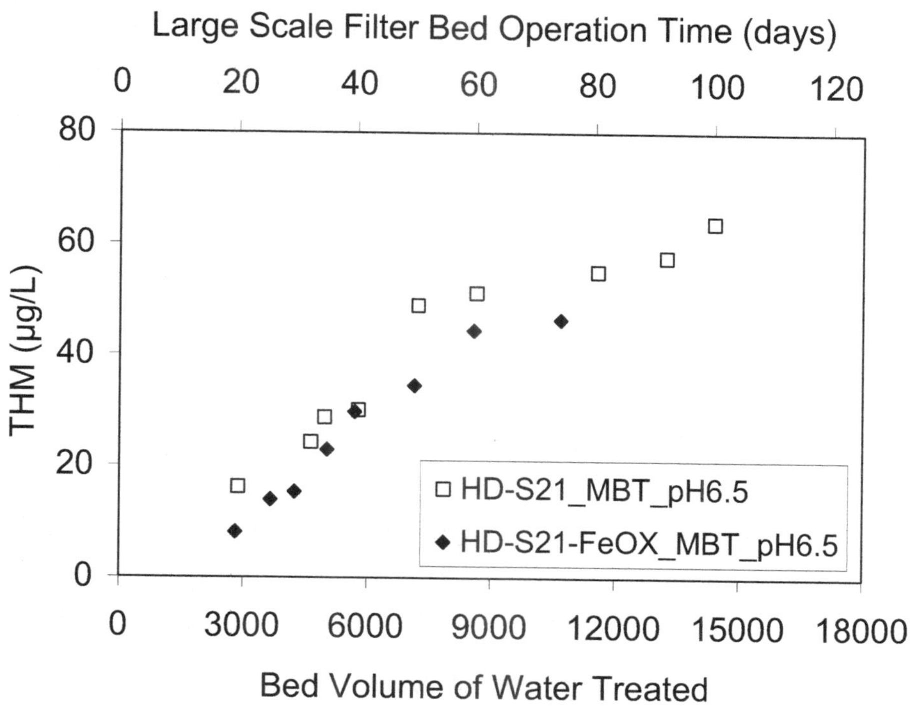

Figure D.9 Myrtle Beach treated water THM formation breakthrough profiles for HD-S21 and HD-S21-FeOX

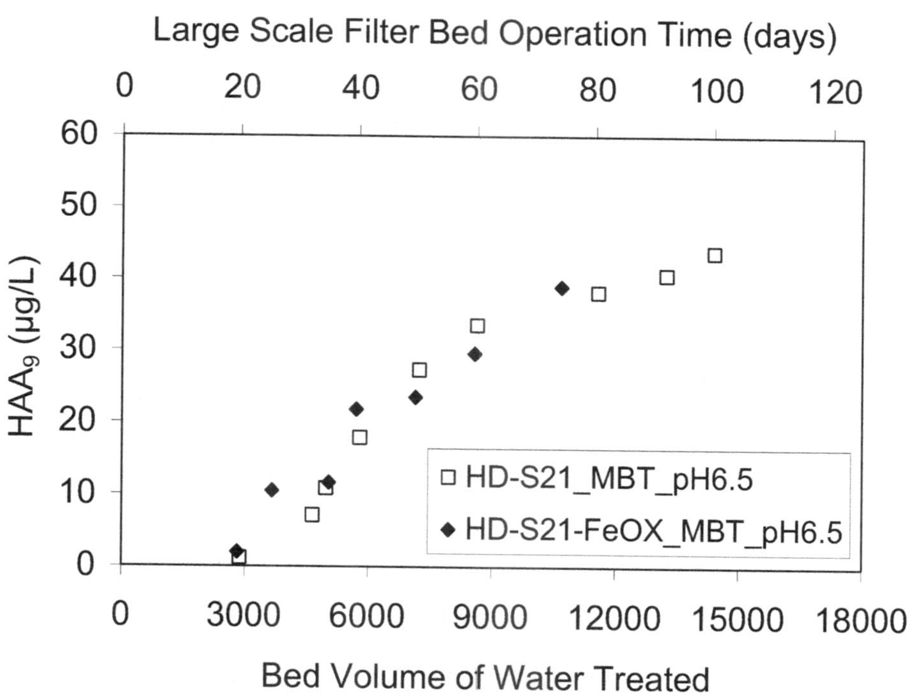

Figure D.10 Myrtle Beach treated water HAA_9 formation breakthrough profiles for HD-S21 and HD-S21-FeOX

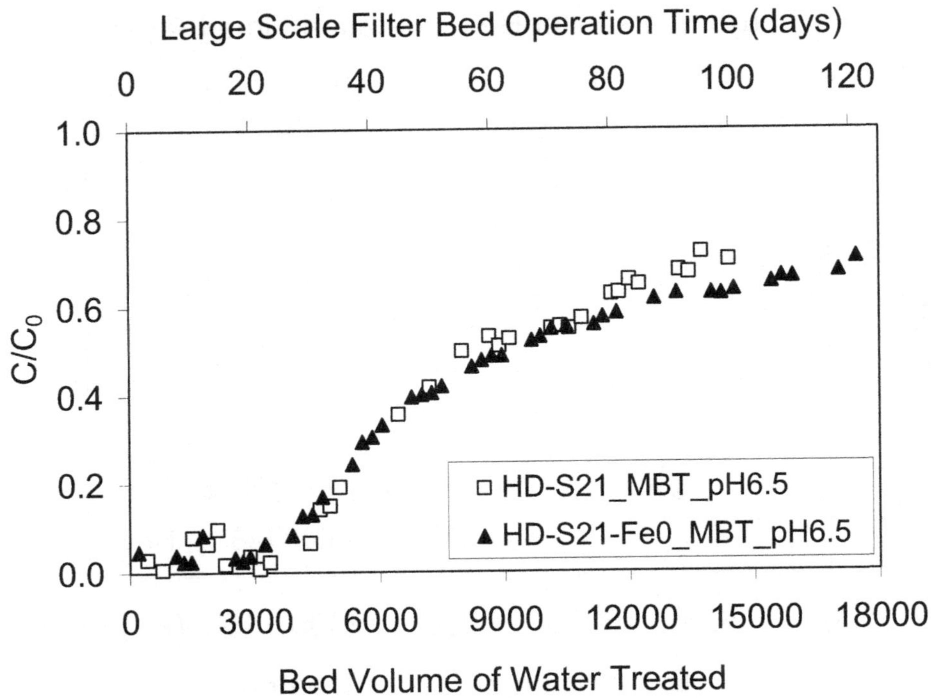

Figure D.11 Myrtle Beach treated water UV_{254} breakthrough profiles for HD-S21 and HD-S21-Fe^0

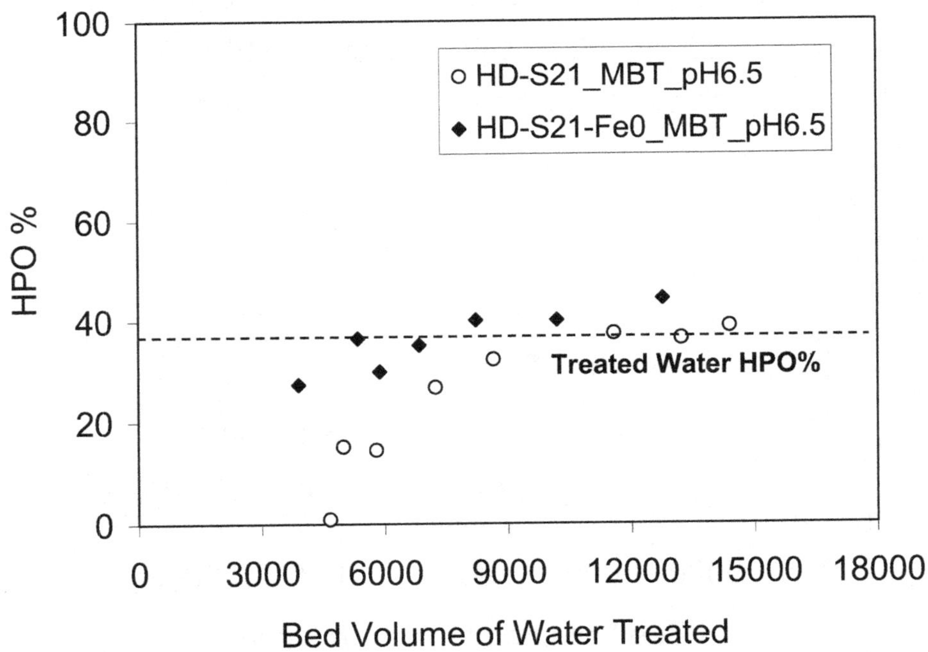

Figure D.12 Myrtle Beach treated water HPO% breakthrough profiles for HD-S21 and HD-S21-Fe^0

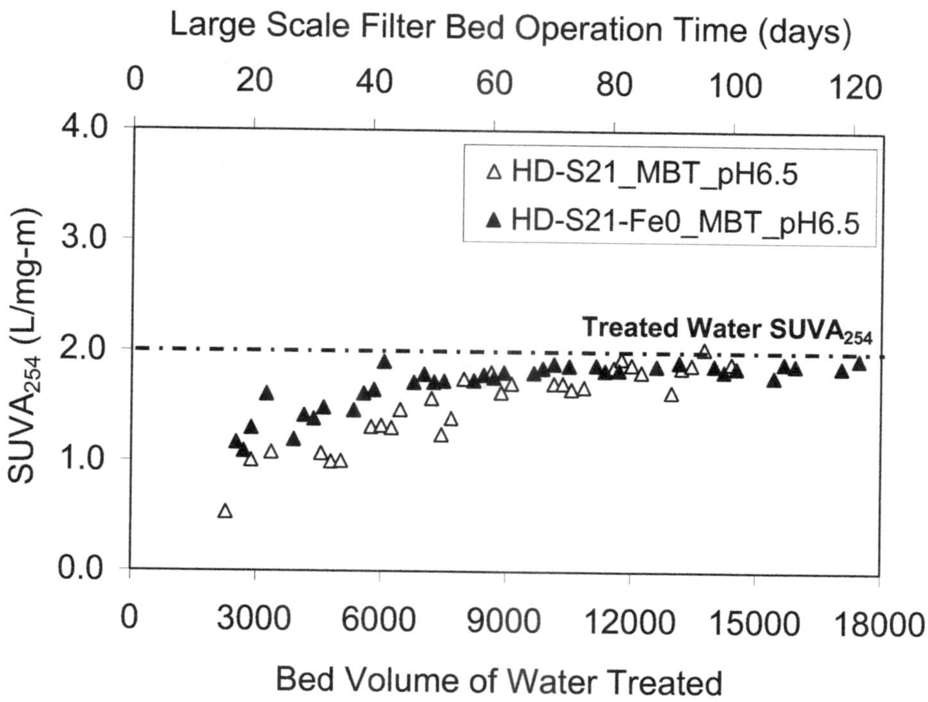

Figure D.13 Myrtle Beach treated water $SUVA_{254}$ breakthrough profiles for HD-S21 and HD-S21-Fe^0

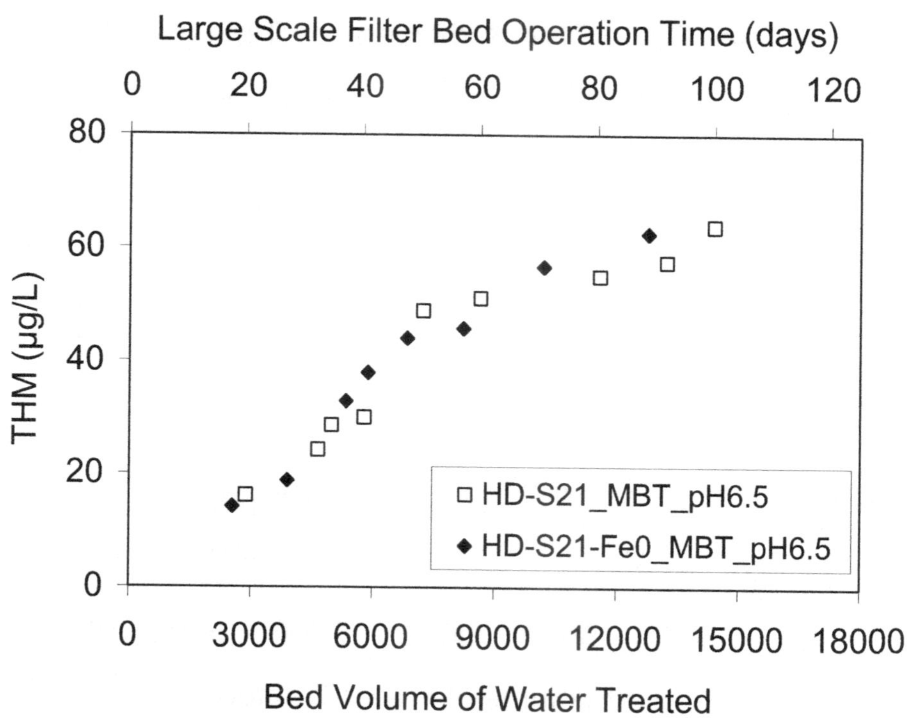

Figure D.14 Myrtle Beach treated water THM formation breakthrough profiles for HD-S21 and HD-S21-Fe^0

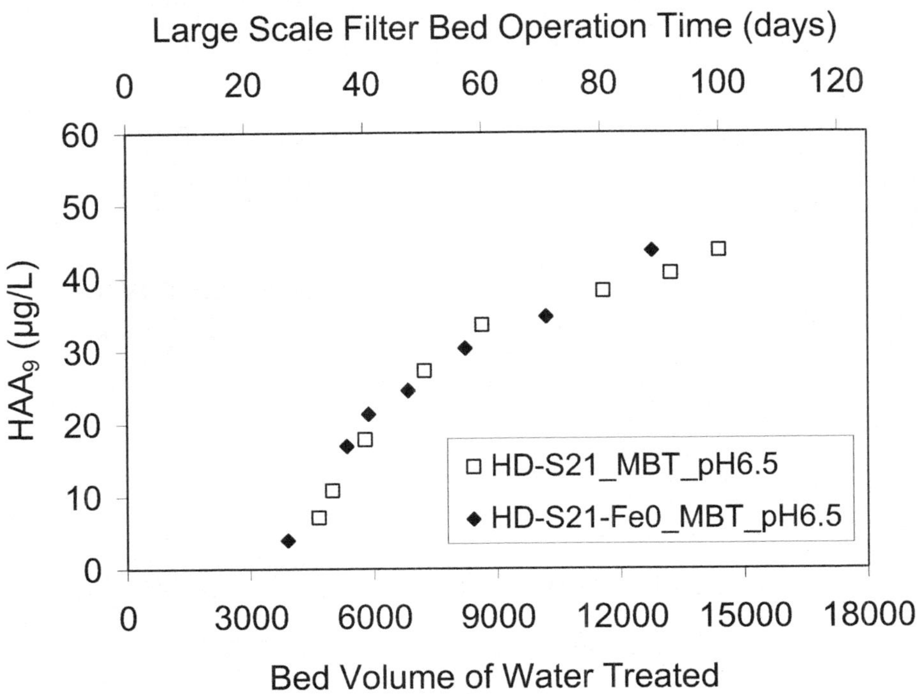

Figure D.15 Myrtle Beach treated water HAA_9 formation breakthrough profiles for HD-S21 and HD-S21-Fe^0

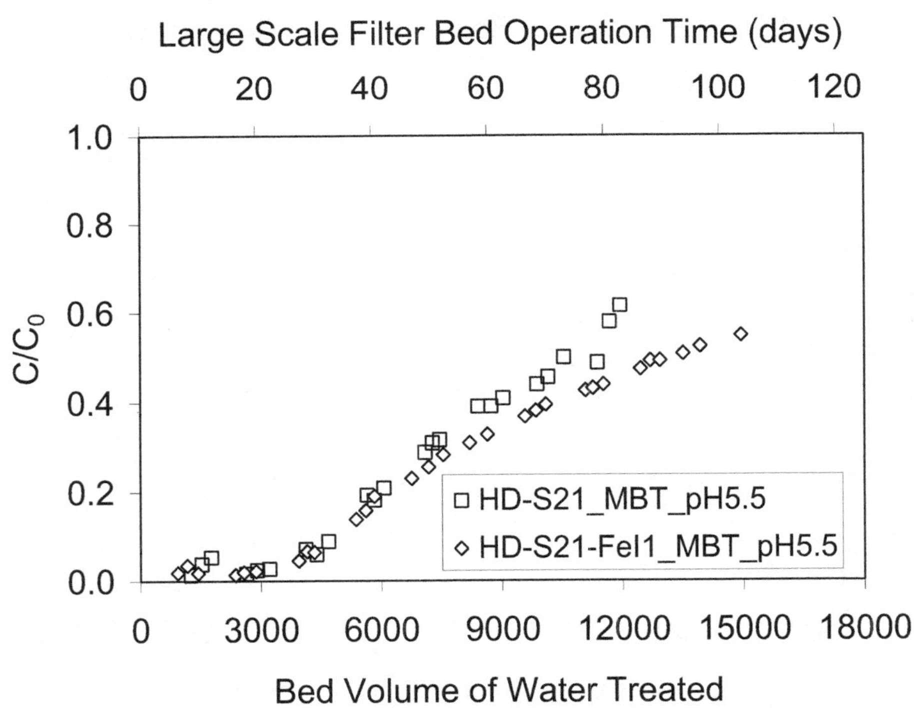

Figure D.16 Myrtle Beach treated water DOM UV_{254} breakthrough profiles for HD-S21 and HD-S21-FeI1 at pH 5.5

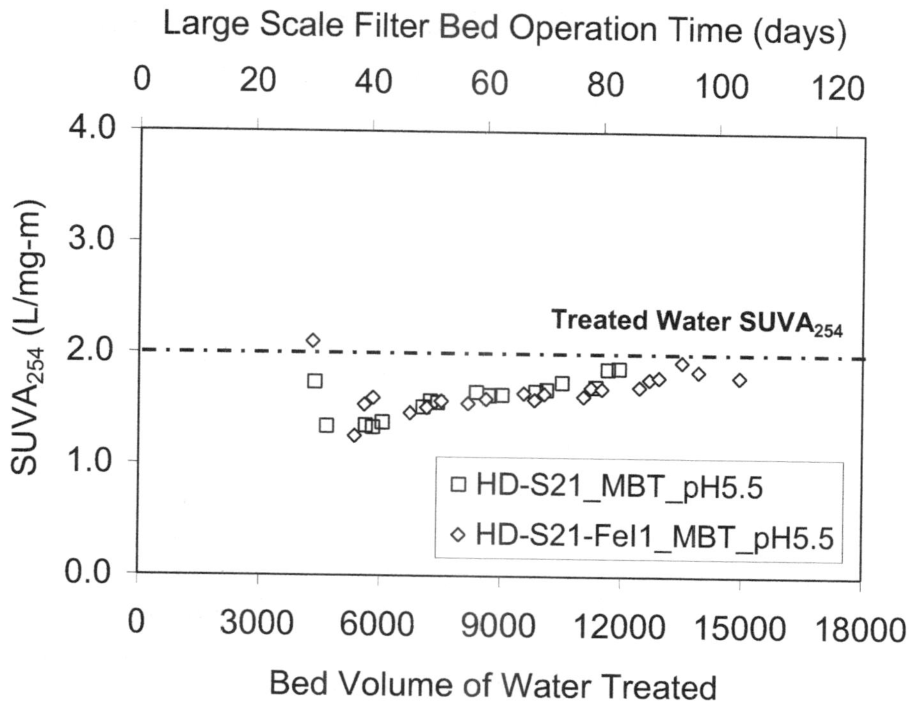

Figure D.17 Myrtle Beach treated water DOM $SUVA_{254}$ breakthrough profiles for HD-S21 and HD-S21-FeI1 at pH 5.5

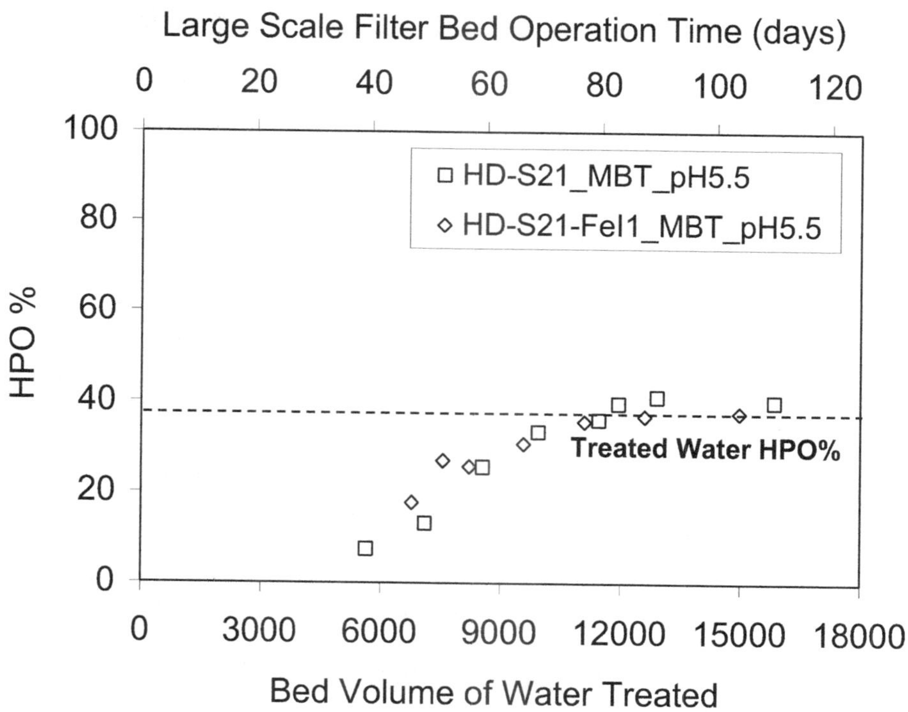

Figure D.18 Myrtle Beach treated water DOM HPO% breakthrough profiles for HD-S21 and HD-S21-FeI1 at pH 5.5

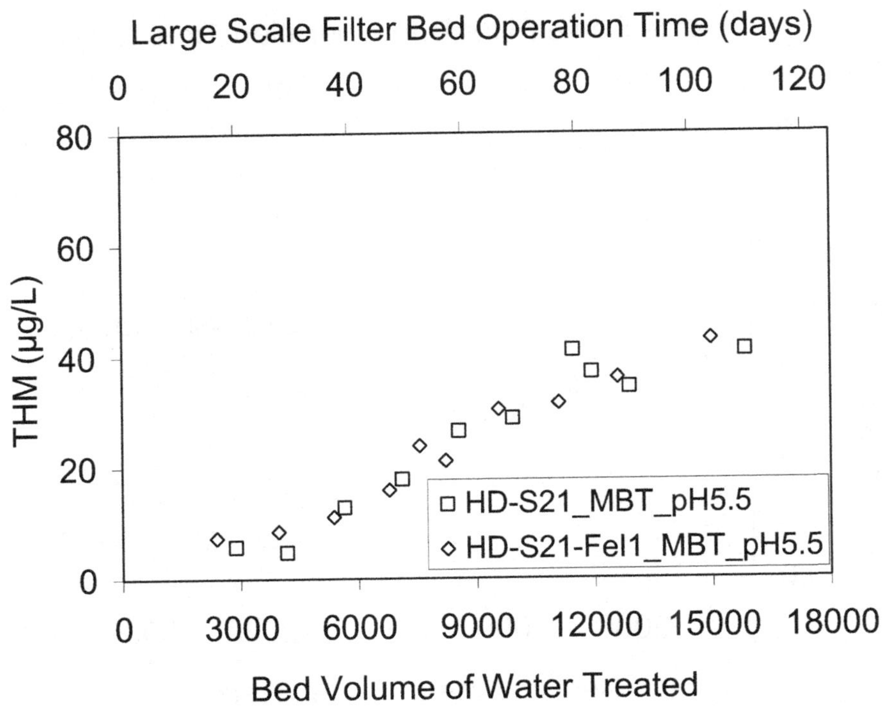

Figure D.19 Myrtle Beach treated water THM formation breakthrough profiles for HD-S21 and HD-S21-FeI1 at pH 5.5

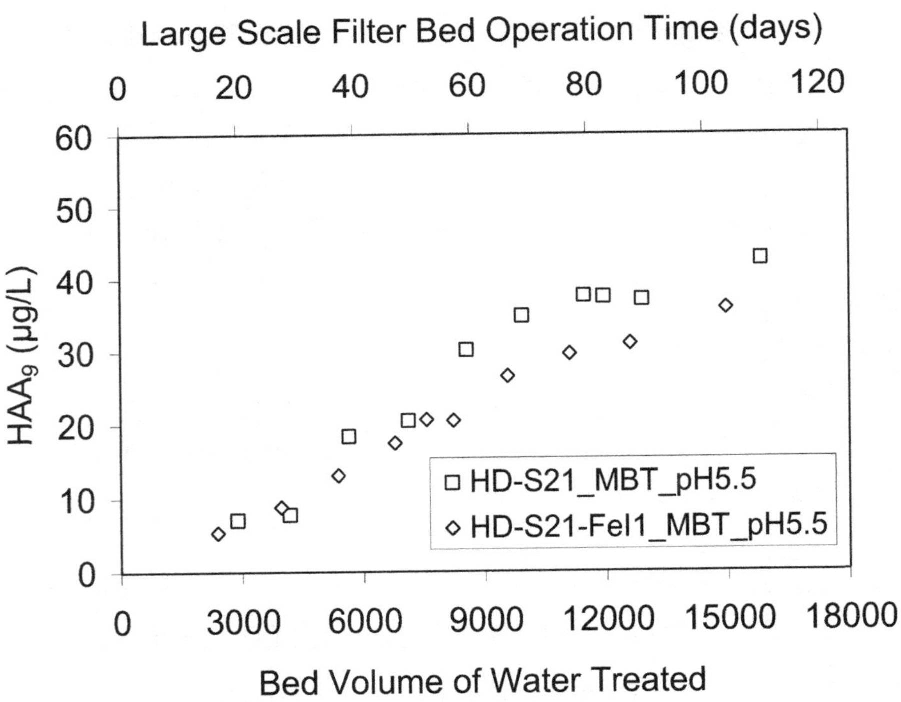

Figure D.20 Myrtle Beach treated water HAA_9 formation breakthrough profiles for HD-S21 and HD-S21-FeI1 at pH 5.5

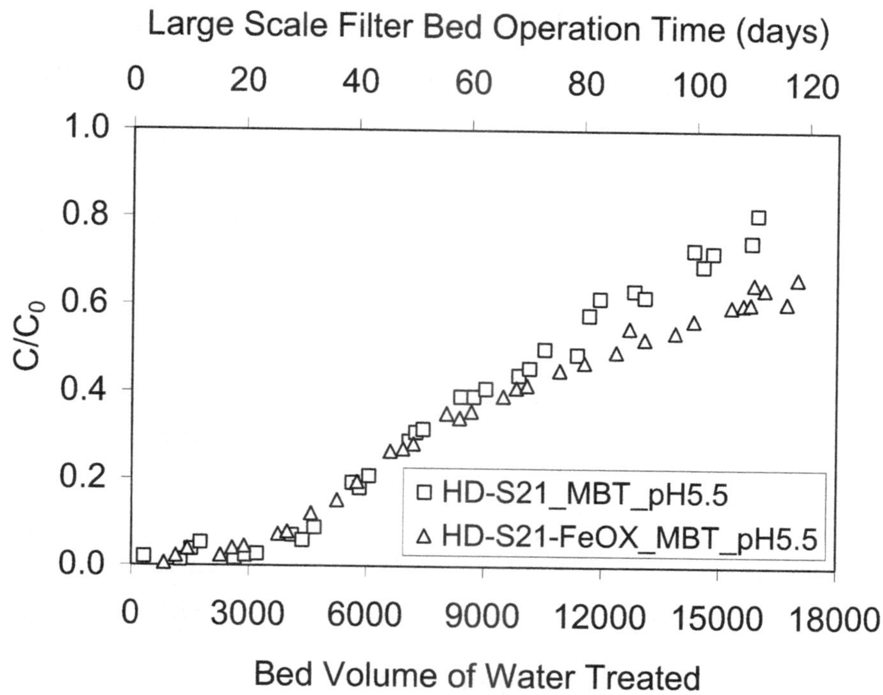

Figure D.21 Myrtle Beach treated water UV_{254} breakthrough profiles for HD-S21 and HD-S21-FeOX at pH 5.5

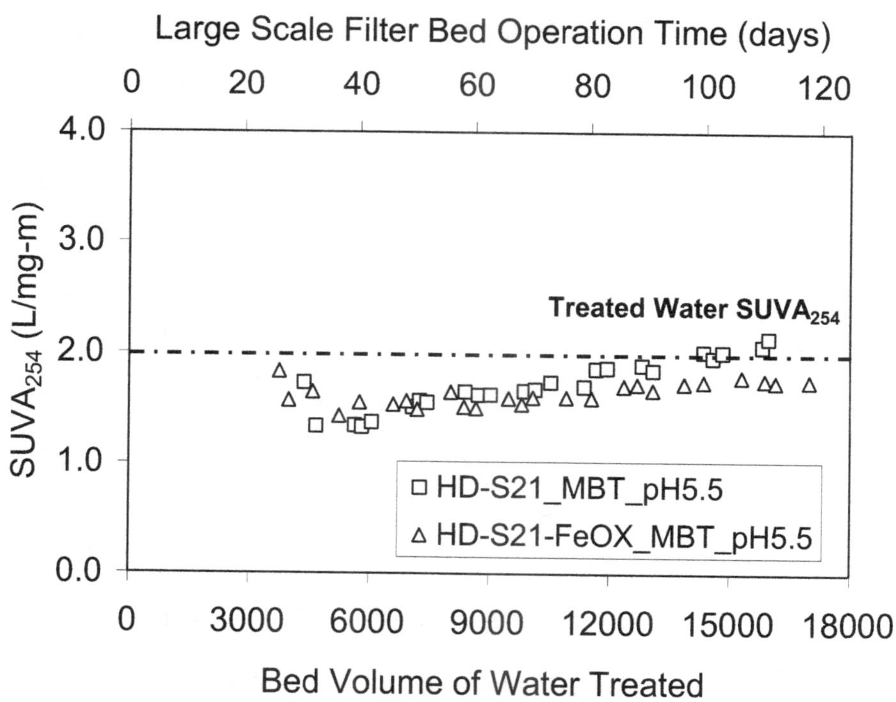

Figure D.22 Myrtle Beach treated water $SUVA_{254}$ breakthrough profiles for HD-S21 and HD-S21-FeOX at pH 5.5

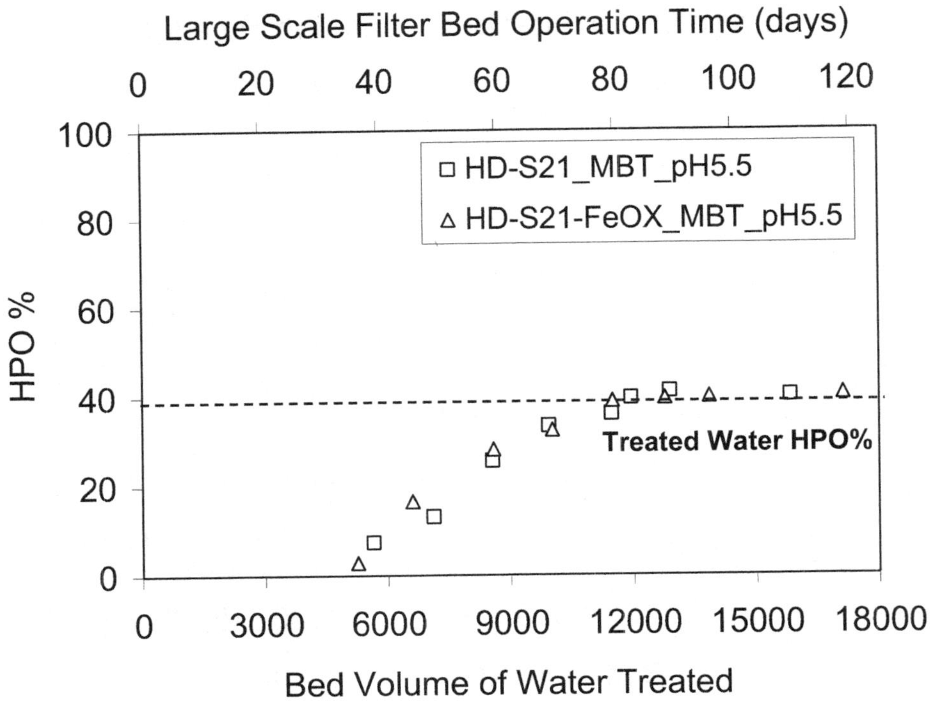

Figure D.23 Myrtle Beach treated water HPO% breakthrough profiles for HD-S21 and HD-S21-FeOX at pH 5.5

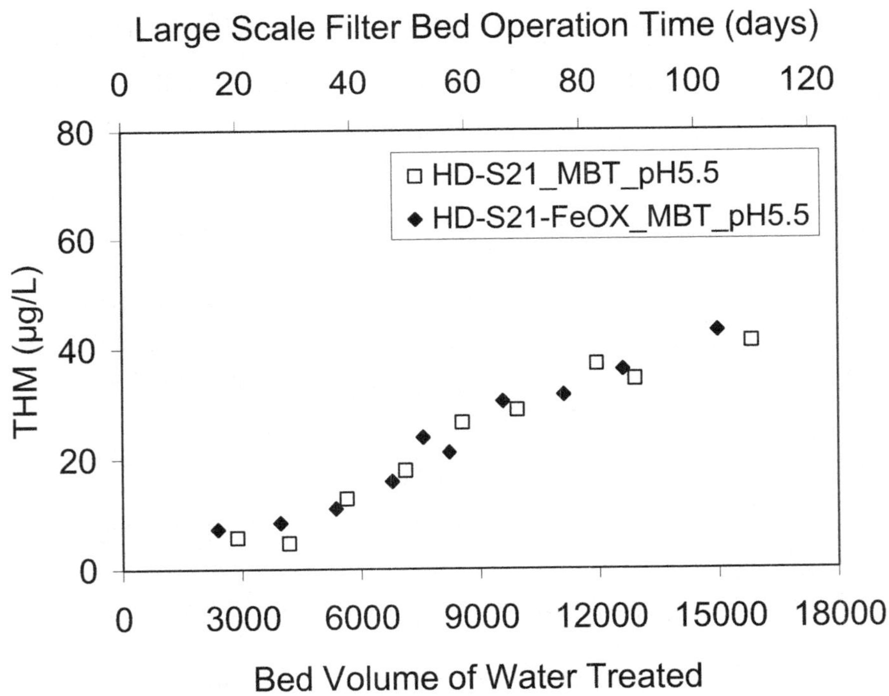

Figure D.24 Myrtle Beach treated water THM formation breakthrough profiles for HD-S21 and HD-S21-FeOX at pH 5.5

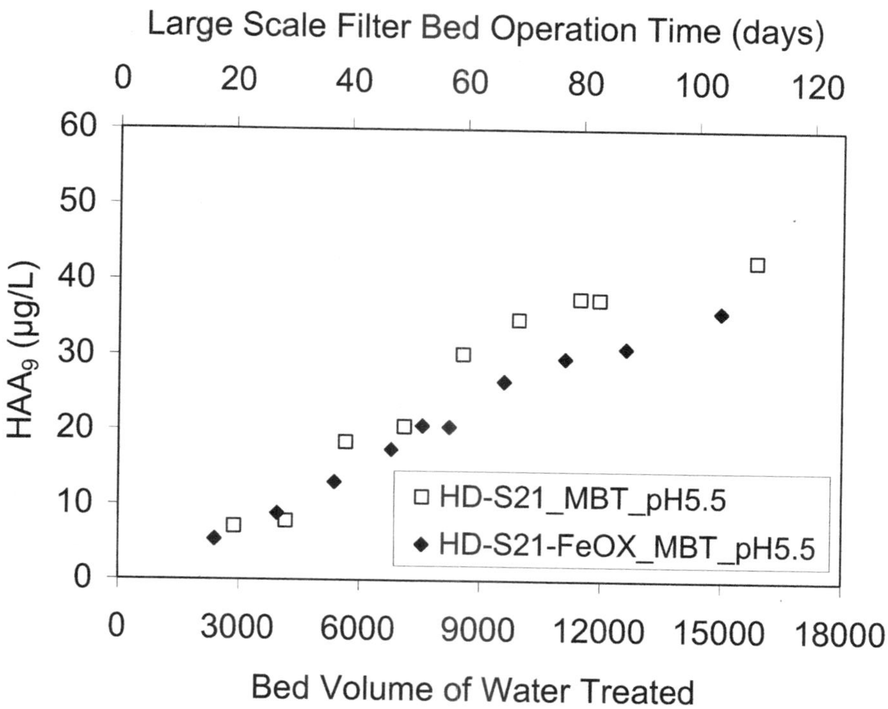

Figure D.25 Myrtle Beach treated water HAA_9 formation breakthrough profiles for HD-S21 and HD-S21-FeOX at pH 5.5

REFERENCES

Abe M., Kawashima K., Kozawa K., Sakai H., Kaneko K. (2000) Amination of activated carbon and adsorption characteristics of its aminated surface. *Langmuir* 16, 5059-5063.

Aika K.I., Hori H., Ozaki A. (1972) Activation of nitrogen by alkali metal promoted transition metal I. Ammonia synthesis over ruthenium promoted by alkali metal. *Journal of Catalysis* 27, 424-431.

Aiken G.R. (1985) Isolation and concentration techniques for aquatic humic substances. In: *Humic Substances in Soil, Sediment, and Water: Geochemistry, Isolation, and Characterization*. Eds., Aiken G.R., McKnight D.M., Wershaw R.L., MacCarthy P., John Wiley & Sons, New York.

Aiken G.R., Malcolm R.L. (1987) Molecular weight of aquatic fulvic acids by vapor pressure osmometry. *Geochim Cosmochim Acta* 51, 2177-2184.

Aiken G.R., Cotsaris E. (1995) Soil and hydrology: Their effect on NOM. *Journal of AWWA* 87(1), 36-45.

Aksoylu A.E., Madalena M., Freitas A., Pereira M.F.R., Figueiredo J.L. (2001) The effects of different activated carbon supports and support modifications on the properties of Pt/AC catalysts. *Carbon* 39(2), 175-185.

Amy G.L., Collins M.R., Kuo C.J., King P.H. (1987) Comparing gel permeation chromatography and ultrafiltration for the molecular weight characterization of aquatic organic matter. *Journal of AWWA* 79(1), 43-49.

Amy G.L., Sierka R.A., Bedessem J., Price D., Tan L. (1992) Molecular size distributions of dissolved organic matter. *Journal of AWWA* 84(6), 67-75.

Andrews S.A. (1993) Organic by-product formation from the ozonation and chlorination of aquatic natural organic matter. *Ph.D. Dissertation*. Dept. of Civil Engineering, University of Alberta, Alberta, Canada.

Baker F.S., Miller C.E., Repik A.J., Tolles E.D. (1992) Activated Carbon. *Kirk-Othmer Encyclopedia of Chemical Technology* 4, 1015-1037.

Barnes P.A., Dawson E.A., Midgley G. (1992) Activation of novel copper carbon catalysts using temperature-programmed mass-spectrometry. *Journal of the Chemical Society-Faraday Transactions* 88(3), 349-353.

Barnes P.A., Dawson E.A. (1994) A thermoanalytical study of the formation and activation of novel copper-carbon catalysts. *Journal of Thermal An*alysis 41(2-3), 621-634.

Barnes P.A., Dawson E.A. (1995) in *Preparation of Catalysts VI*. Poncelet et al. eds., Elsevier, Amsterdam.

Baylis J.R. (1929) The activated carbons and their use in removing objectionable taste and odors from water. *Journal of AWWA* 21, 787-814.

Bjelopavlic M., Newcombe G., Hayes R. (1999) Adsorption of NOM onto activated carbon: Effect of surface charge, ionic strength, and pore volume distribution. *Journal of Colloid and Interface Science* 210, 271-280.

Boehm H.P. (1974) Funktionelle Gruppen an Festkőrperoberflächen. *Chemie-Ingenieur Technik* 46(17), 716-719.

Bolto B., Dixon D., Eldridge R., King S. (2001) Cationic polymer and clay or metal oxide combinations for natural organic matter removal. *Water Research* 35, 2669-2676.

Bourke M., Slunjski M. (1999) MIEX TOC process launched in Western Australia. A new Australian development. *Water* (Artarmon, Aust.) 26(6), 17-18.

Budd G.C., Long B.W., Edwards-Brandt C. (2005) *Evaluation of MIEX process impacts on different source waters*. American Water Works Association Research Foundation Report.

Cameron R.S., Swift R.S., Thornton B.K., Posner M.A. (1972) Calibration of gel permeation chromatography materials for use with humic acid. *Journal of Soil Science* 23(3), 342-349.

Cameron R.S., Thornton B.K., Swift R.S., Posner A.M. (1972) Molecular weight and shape of humic acid from sedimentation and diffusion measurements on fractionated extracts. *Journal of Soil Science* 23, 394.

Cheng W., Dastgheib S.A., Karanfil T. (2005) Adsorption of dissolved natural organic matter by modified activated carbons. *Water Research* 39, 2281-2290. Elsevier.

Chiang H.L., Huang C.P., Chiang P.C., You J.H. (1999) Effect of metal additives on the physico-chemical characteristics of activated carbon exemplified by benzene and acetic acid adsorption. *Carbon* 37, 1919-1928.

Chin Y.P., Aiken G., O'Loughlin E. (1994) Molecuilar weight, polydispersity, and spectroscopic properties of aquatic humic substances. *Environmental Science & Technology* 28, 1853-1858.

Conte P., Piccolo A. (1999) Conformational arrangement of dissolved humic substances. Influence of solution composition on association of humic molecules. *Environmental Science & Technology* 33(10), 1682-1690.

Corapcioglu M.O., Huang C.P. (1987) The surface acidity and characterization of some commercial activated carbon. *Carbon* 25, 569-579.

Cornel P.K., Summers R.S., Roberts P.V. (1986) Diffusion of humic acid in aqueous solution. *Journal of Colloid and Interface Science* 110, 149-164.

Cornell R.M., Schwertmann U. (1996) *The Iron Oxides: Structure, Properties, Reactions, Occurrence and Uses.* VCH Verlagsgesellschaft Weinheim, Germany.

Coughlin R.W., Ezra F.S. (1968) Role of surface acidity in the adsorption of organic pollutants on the surface of carbon. *Environmental Science & Technology* 2(4), 291-297.

Cronce D.T., Mansour A.N., Brown R.P., Beard B.C. (1997) SeF_4 and SF_4 fluorination of BPL activated carbon surfaces. *Carbon* 35(4), 483-495.

Crozes G., White P., Marshall M. (1995) Enhanced coagulation: Its effect on DOM removal and chemical costs. *Journal of AWWA* 87(1), 79-89.

Culp R.L., Clark R.M. (1983) Granular activated carbon installations. *Journal of AWWA* 75(8), 398-405.

Dastgheib S.A., Rockstraw D.A. (2001) Pecan shell activated carbon: Synthesis, characterization, and application for the removal of copper from aqueous solution. *Carbon* 39, 1849-1855.

Dastgheib S.A., Karanfil T., Cheng W. (2004) Tailoring activated carbons for enhanced removal of natural organic matter from natural waters. *Carbon* 42, 547-557.

Dubinin M.M., Zaverina E.D. (1949) Structure and adsorptive properties of zinc chloride-activated carbon. *Doklady Akademii Nauk SSSR* 68, 91-94.

Economy J., Foster K.L., Andreopoulos A.G., Jung J. (1992) Tailoring carbon fibers for adsorbing volatiles. *Chemtech* 22(10), 597-603.

Edwards M. (1997) Predicting TOC removal during enhanced coagulation. *Journal of AWWA* 89(5), 78-89.

Edzwald J.K., van Benschoten J.E. (1990). Aluminum coagulation of natural organic matter. In: *Chemical Water and Wastewater Treatment*. Eds. Hahn H.H., Klute R. Springer-Verlag, Berlin.

Ermolenko I.N., Lyubliner I.P., Gulko N.V. (1990) *Chemically Modified Carbon Fibers and Their Applications*. Translated by Titovets E.P. VCH Publishers, New York.

Figueiredo J.L., Pereira M.F.R., Freitas M.M.A., Orfao J.J.M. (1999) Modification of the surface chemistry of activated carbons. *Carbon* 37, 1379-1389.

Freese S.D., Nozaic D.J., Pryor M.J., Rajogopaul R., Trollip D.L., Smith R.A. (2001) Enhanced coagulation: A viable option to advance treatment technologies in the South African context. *Water Science and Technology: Water Supply* 1(1), 33-41.

Garten V.A., Weiss D.E. (1957) A new interpretation of the acidic and basic structures in carbons II. The chromene-carbonium ion couple in carbon. *Austral. J. Chem*. 10, 309.

Ghosh K., Schnitzer M. (1980) Macromolecular structures of humic substances. *Soil Science* 129, 266-276.

Gibeson R.C., Walker J.P. (1966) Recreation of nuclear graphite with water vapor. Part I. Effect of hydrogen and water vapor partial pressures. *Carbon* 3, 521-525.

Glaze W.H. (1987) Dinking water treatment with ozone. *Environmental Science & Technology* 21(3), 224-230.

Glaze W.H. (1990) Group report: "What is the composition of organic acids in aquatic systems and how are they characterized?" In: *Organic Acids in Aquatic Ecosystems*. Eds., Perdue E.M., Gjessing E.T. John Wiley & Songs, New York.

Gomez-Serrano V., Acedo-Ramos M., Lopez-Peinado A.J., Valenzuela-Calahorro C. (1997) Mass and surface changes of activated carbon treated with nitric acid. Thermal behavior of the samples. *Thermochimica Acta* 291(1-2), 109-115.

Gregg S.J., Sing K.S.W. (1982) *Adsorption, Surface Area and Porosity*. Academic Press, New York.

Gu B., Schmitt J., Chen Z., Liang L., McCarthy J.F. (1995) Adsorption and desorption of different organic matter fractions on iron oxide. *Geochimica Cosmochimica Acta* 59(2), 219-229.

Hansen R.S., Hansen R.D., Craig R.P. (1953) The effect of steam activation on multilayer adsorption from solution by carbon. *Journal of Physical Chemistry* 57, 215-219.

Hashimoto K., Miura K., Yoshikawa F., Imai I. (1979) Change in pore structure of carbonaceous materials during activation and adsorption performance of activated carbon. *Industrial and Engineering Chemistry Process Design and Development.* 18(1), 72-80.

Henning G.R. (1962) Catalytic oxidation of graphite. *Journal of Inorganic and Nuclear Chemistry* 24, 1129-1137.

Hooper S.M., Summers R.S., Solarik G., Owen D.M. (1996) Improving GAC Performance by Optimized Coagulation. *Journal of AWWA* 88 (8), 107-119.

Huff C.B. et al. (1965) Study of ultraviolet disinfection of water and factors in treatment efficiency. *Public Health Reports* 80(8), 695-705.

Hur J. (2002) Effects of Humic Substances (HS) Heterogeneity on Pyrene Binding to Dissolved and Mineral-Bound HS. Ph.D. Thesis. Clemson University, Clemson, South Carolina.

Hutton H.D., Pocard N.L., Alsmeyer D.C., Schueller O.J.A., Spontak R.J., Huston M.E., Huang W., McCreery R.L., Neenan T.X., Callstrom M.R. (1993) Preparation of nanoscale platinum(0) clusters in glassy-carbon and their catalytic activity. *Chemistry of Materials* 5(12), 1727-1738.

Jacangelo J.G., DeMarco J., Owen D.M., Randtke S.J. (1995) Selected processes for removing NOM: an overview. *Journal of AWWA* 87(1), 64-77.

Jansen R.J.J., van Bekkum H. (1994) Amination and ammoxidation of activated carbons. *Carbon* 32(8), 1507-1516.

Jansen R.J.J., van Bekkum H. (1995) XPS of nitrogen-containing functional groups on activated carbon. *Carbon* 33(8), 1021-1027.

Johnstone H.F., Chen C.Y., Scott D.S. (1952) Kinetics of the steam-carbon reaction in porous graphite tubes. *Industrial & Engineering Chemistry Research* 44, 1564-1569.

Kaneko K., Nakahigashi Y., Nagata K. (1988) Microporosity and adsorption characteristics against NO, SO_2 and NH_3 of pitch-based activated carbon fibers. *Carbon* 26, 27-32.

Kaneko K., Yang C.M., Ohkubo T., Kimura T., Iiyama T., Touhara H. (2000) Cluster mediated filling of water and alcohol on microporous carbon alloys. *Second Pacific Basin Conference on Adsorption Science and Technology*, Brisbane, Australia, 2000:1-5.

Karanfil T., Schlautman M.A., Weber W.J., Jr. (1994) Impacts of dissolved-oxygen on the sorption of humic substances and the subsequent inhibition of o-cresol uptake by granular activated carbon. *Water Research* 28 (7), 1673-1678.

Karanfil T., Kilduff J.E., Schlautman M.A., Weber W.J., Jr. (1996a) Adsorption of organic macromolecules by granular activated carbon. 1. Influence of molecular properties under anoxic solution conditions. *Environmental Science & Technology* 30, 2187–2194.

Karanfil T., Schlautman M.A., Kilduff J.E., Weber W.J., Jr. (1996b) Adsorption of organic macromolecules by granular activated carbon. 2. Influence of dissolved oxygen. *Environmental Science & Technology* 30, 2195–2201.

Karanfil T., Kilduff J.E., Schlautman M.A., Weber W.J., Jr. (1998) The oxygen sensitivity of organic macromolecule sorption by activated carbon: Effects of solution chemistry. *Water Research* 32, 154–164.

Karanfil T., Kilduff J.E. (1999) The role of GAC surface chemistry on the adsorption of organic compounds: 1. Priority pollutants. *Environmental Science & Technology* 33, 3217-3224.

Karanfil T., Kilduff J.E., Kitis M., Wigton A. (1999) Role of granular activated carbon surface chemistry on the adsorption of organic compounds: 2. Natural organic matter. *Environmental Science & Technology* 33, 3225-3233.

Karanfil T., Erdogan I., Schlautman M.A. (2003) Selecting filter membranes for measuring TOC and UV_{254}. *Journal of AWWA* 95(3), 86-100.

Karanfil T., Erdogan I., Schlautman M. (2005) The impact of filtrate turbidity on UV_{254} and $SUVA_{254}$ determinations. *Journal of AWWA* 97 (5), 125-136.

Karanfil T., Hong, Y., Song, H., Orr, O. (2006) *Exploring the pathways of HAA formation during chloramination.* Denver, Colo.: AwwaRF and AWWA.

Karanfil T. (2006) Activated carbon adsorption in drinking water treatment. In: *Activated Carbon Surfaces in Environmental Remediation*. Ed. Bandosz T.J. Academic Press, New York.

Kilduff J.E., Karanfil T., Chin Y.P., Weber W.J., Jr. (1996) Adsorption of natural organic polyelectrolytes by activated carbon: A high-performance size exclusion chromatography study. *Environmental Science & Technology* 30, 1336-1343.

Kilduff J.E., Mattaraj S., Wigton A., Kitis M., Karanfil T. (2004) Effects of reverse osmosis isolation on reactivity of naturally occurring dissolved organic matter in physicochemical processes . *Water Research* 38(4), 1026-1036.

King A. (1937) Studies in chemisorption on charcoal: The influence of temperature of activation on the sorption of acids and bases. *Journal of the Chemical Society* 2, 1489-1491.

Kitis M., Karanfil T., Wigton A., Kilduff J.E. (2002) Probing reactivity of dissolved organic matter for disinfection by-product formation using XAD-8 resin adsorption and ultrafiltration fractionation. *Water Research* 36(15), 3834-3848.

Korshin G.V., Benjamin M.M., Sletten R.S. (1997) Adsorption of natural organic on iron oxide: effect on NOM composition and formation of organohalide compounds during chlorination. *Water Research* 31(7), 1643-1650.

Krasner S.W., Croue J.P., Buffle J., Perdue E.M. (1996) Three approaches for characterizing NOM. *Journal of AWWA* 88(6), 66-79.

Krczil F. (1937) Production of activated carbon. *US Patent* #2,083,303.

Krebs C., Smith J.M. (1985) Kinetics of steam regeneration of virgin and phenol-loaded activated carbon. *Carbon* 23(2), 223-235.

Kutics K., Suzuki M. (1993) Adsorption of organics on surface treated activated carbon fibers. *Hungarian Journal of Industrial Chemistry*. 21, 93-99.

Lafrance P., Mazet M. (1989) Adsorption of humic substances in the presence of sodium salts. *Journal of AWWA* 81(4), 155-162.

Lambiotte A. (1942) Process of continuous carbonation of cellulosic materials. *US Patent* #2,289,917.

Lee M.C., Snoeyink V.L., Crittenden J.C. (1981) Activated carbon adsorption of humic substances. *Journal of AWWA* 73, 440–446.

Leenheer J.A., Croue J.P. (2003) Characterizing aquatic dissolved organic matter. *Environmental Science & Technology* 37(1), 19-26.

Leon C.A., Radovic L.R. (1994) In *Chemistry and Physics of Carbon*, Ed. P.A. Thrower, Vol. 24, New York: Marcel Dekker.

Letterman R.D., Amirtharajah A., O'Melia C.R. (1999) Chapter 6: Coagulation and Flocculation. In: *Water Quality & Treatment*. Ed., Letterman R.D. American Water Works Association. McGraw-Hill Inc., New York.

Li F., Yuasa A., Ebie K., Zuma Y. (2003) Microcolumn test and model analysis of activated carbon adsorption of dissolved organic matter after precoagulation: Effects of pH and pore size distribution. *Journal of Colloid and Interface Science* 262, 331-341.

Liang L., Singer P.C. (2003) Factors influencing the formation and relative distribution of haloacetic acids and trihalomethanes in drinking water. *Environmental Science & Technology* 37(13), 2920-2928.

Lin Y.R., Teng H. (2002) Mesoporous carbons from waste tire char and their application in wastewater discoloration. *Microporous and Mesoporous Materials*. 54, 167-174.

Louwers S.P.A., Prins R. (1992) Ni EXAFS studies of the Ni-Mo-S structure in carbon-supported and alumina-supported Ni-Mo catalysts. *Journal of Catalysis* 133(1), 94-111.

Malcolm R.L. (1985) Geochemistry of stream fulvic and humic substances. In: *Humic Substances in Soil, Sediment, and Water: Geochemistry, Isolation, and Characterization*. Eds., Aiken G.R., McKnight D.M., Wershaw R.L., MacCarthy P. John Wiley & Sons, New York.

Mangun C.L., Benak K.R., Daley M.A., Economy J. (1999a) Oxidation of activated carbon fibers: Effect on pore size, surface chemistry, and adsorption properties. *Chemistry of Materials* 11, 3476-3483.

Mangun C.L., Braatz R.D., Economy J., Hall A.J. (1999b) Fixed bed adsorption of acetone and ammonia onto oxidized activated carbon fibers. *Industrial & Engineering Chemistry Research* 38, 3499-3504.

Mangun C.L., Benak K.R., Daley M.A., Economy J., Foster K.L. (2001) Surface chemistry, pore sizes and adsorption properties of activated carbon fibers and precursors treated with ammonia. *Carbon* 39, 1809-1820.

MacCarthy P., Malcolm R.L., Clapp C.E., Bloom P.R. (1990) An introduction to soil humic substances, In: *Humic Substances in Soil and Crop Sciences: Selected Readings*. Eds., Suffet I.H., MacCarthy P. Advances in Chemistry series 219. American Chemical Society, Washington DC.

Maroto-Valer M.M., Dranca I., Lupascu T., Nastas R. (2004) Effect of adsorbate polarity on thermodesortion profiles from oxidized and metal-impregnated activated carbons. *Carbon* 42, 2655-2659.

Matilainen A., Lindqvist N., Korhonen S., Tuhkanen T. (2002) Removal of NOM in the different stages of the water treatment process. *Environmental International* 28, 457-465.

McCreary J.J., Snoeyink V.L. (1980) Characterization and activated carbon adsorption of several humic substances. *Water Research* 14, 151-160.

McGuire, M.J., McLain, J.L., Obolensky, A. (2002) *Information collection rule data analysis*. American Water Works Association Research Foundation Report.

Menendez J.A., Illan-Gomez M.J., Leon C.A., Radovic L.R. (1995) On the difference between the isoelectric point and the point of zero charge of carbons. *Carbon* 33(11), 1655-1659.

Menendez J.A., Phillips J., Xia B., Radovic L.R. (1996a) On the modification and characterization of chemical surface properties of activated carbon: In the search of carbons with stable surface properties. *Langmuir* 12, 4404-4410.

Menendez J.A., Radovic L.R., Xia B., Phillips J.J. (1996b) Low-temperature generation of basic carbon surfaces by hydrogen spillover. *Journal of Physical Chemistry* 100, 17243-17248.

Miyazaki A., Shibazaki K., Nakano Y., Ogawa M., Balint I. (2004) Efficient catalytic reduction of concentrated nitric acid on the adsorption sites of activated carbon. *Chemistry Letters* 33(4), 418-419.

Mochida I., Kawano S. (1991) Capture of ammonia by active carbon fibers further activated with sulfuric acid. *Industrial & Engineering Chemistry Research* 30, 2322-2327.

Montes-Moran M.A., Suarez D., Menendez J.A., Fuente E. (2004) On the nature of basic sites on carbon surfaces: An overview. *Carbon* 42 (7), 1219-1225.

Moore B.C., Cannon F.S., Westrick J.A., Metz D.H., Shrive C.A., DeMarco J., Hartman D.J. (2001) Changes in GAC pore structure during full-scale water treatment at Cincinnati: A comparison between virgin and thermally reactivated GAC. *Carbon* 39, 789-807.

Moreno-Castilla C., Carrasco-Marin F., Mueden A. (1997) The creation of acid carbon surface by treatment with $(NH_4)_2S_2O_8$. *Carbon* 35(10-11), 1619-1626.

Newcombe G., Drikas M. (1997) Adsorption of NOM onto activated carbon: Electrostatic and non-electrostatic effects. *Carbon* 35(9), 1239–1250.

Newcombe G., Drikas M., Assemi A., Beckett R. (1997a) Influence of characterized natural organic material on activated carbon adsorption: I. Characterization of concentrated reservoir water. *Water Research* 31(5), 965-972.

Newcombe G., Drikas M., Hayes R. (1997b) Influence of characterised natural organic material on activated carbon adsorption: II. Effect on pore volume distribution and adsorption of 2-methylisoborneol.*Water Research* 31, 1065–1073.

Newcombe G. (1999) Charge vs. porosity - Some influences on the adsorption of natural organic matter (NOM) by activated carbon. *Water Science and Technology* 40(9), 191-198.
Noh J.S., Schwarz J.A. (1990) Effect of nitric acid treatment on the surface acidity of activated carbons. *Carbon* 28(5), 675-682.
Nowack K.O., Cannon F.S., Arora H. (1999) Ferric chloride plus GAC for removing TOC. *Journal of AWWA* 91(2), 65-78.
Nowack K.O. (2002) Enhancing activated carbon adsorption of 2-methylisoborneol (MIB). Ph.D. Dissertation. Dept. of Civil and Environmental Engineering, Pennsylvania State University, University Park, PA.
Nowack K.O., Cannon F.S., Mazyck D.W. (2004) Enhancing activated carbon adsorption of 2-methylisoborneol: Methane and steam treatments. *Environmental Science & Technology* 38, 276-284.
O'Loughlin E., Chin Y.P. (2001) Effect of detector wavelength on the determination of the molecular weight of humic substance by high-pressure size exclusion chromatography. *Water Research* 35(1), 333-338.
Osmond N.M. (2000) Activated carbon fibre adsorbent materials. *Adsorption Science & Technology* 18(6), 529-539.
Owen D.M., Amy G.L., Chowdhury Z.K. (1993) *Characterization of natural organic matter and its relationship to treatability*. American Water Works Association Research Foundation Report.
Owen D.M., Amy G.L., Chowdhury Z.K., Paode R., McCoy G., Viscosil K. (1995) NOM characterization and treatbility. *Journal of AWWA* 87(1), 46-63.
Pelekani C., Newcombe G., Snoeyink V.L., Hepplewhite C., Assemi S., Beckett R. (1999) Characterization of natural organic matter using high performance size exclusion chromatography. *Environmental Science & Technology* 33, 2807-2813.
Pelekani C., Snoeyink V.L. (1999) Competitive adsorption in natural water: Role of activated carbon pore size. *Water Research* 33(5), 1209-1219.
Pinna F., Signoretto M., Strukul G., Benedetti A., Malentacchi M., Pernicone N. (1995) Ruthenium as a dispersing agent in carbon-supported palladium. *Journal of Catalysis* 155(1), 166-169.
Pittman C.U., Jr., He G.R., Wu B., Gardner S.D. (1997) Chemical modification of carbon fiber surfaces by nitric acid oxidation followed by reaction with tetraethylenepentamine. *Carbon* 35(3), 317-331.
Pontius F. (1999) Introducing the new federal regulations for disinfectants, DBPs, and microbials. *Opflow* 25 (3), 6-7.
Potoczna-Petru K., Krajczyk L. (1995) Microstructure evolution of Co particles supported on carbon induced by oxidation-reduction treatment *Journal of Materials Science Letters* 14(18), 1294-1297.
Puri B.R. (1970) Surface complexes on carbons. In: *Physics and Chemistry of Carbon* Vol. 6. Ed., Walker Jr P.L. Marcel Dekker, New York.
Puri B.R. (1980) Carbon adsorption of pure compounds and mixtures from solution phase. In: *Activated Carbon Adsorption of Organics from the Aqueous Phase.* Vol. I. Eds., Suffet I.H., McGuire M.J. Ann Arbor Science, Ann Arbor, Michigan.
Radovic L.R., Rodríguez-Reinoso, F. (1997) Carbon Materials in Catalysis. In: *Chemistry and Physics of Carbon*. Vol. 25. Ed., Thrower P.A. Marcel Dekker, New York.

Randtke S.J., Jepsen C.P. (1982) The effects of salts on activated carbon. *Journal of AWWA* 74, 84-93.

Randtke S.J. (1988) Organic contaminant removal by coagulation and related process combinations. *Journal of AWWA* 80, 40-56.

Randtke S.J., Hoehn R.C. (1999) *Removal of DBP Precursors by Enhanced Coagulation and Lime Softening*. American Water Works Association Research Foundation Report.

Rangel-Mendez J.R., Cannon F.S. (2005) Improved activated carbon by thermal treatment in methane and steam: Physicochemical influences on MIB sorption capacity. *Carbon* 43, 467-479.

Reckhow D.A., Singer P.C., Malcolm R.L. (1990) Chlorination of humic materials: By product formation and chemical interpretations. *Environmental Science & Technology* 24(11), 1655-1664.

Reichenberger J. (1974) Zum mechanismus der thermischen regeneration von beladenen activohlen im wirbelbett in gegenwart von wasserdampt, Dissertation, Technische Hochschule, Aachen.

Richard D., Gallezot P., Neibecker D., Tkatchenko I. (1989) Characterization and selectivity in cinnamaldehyde hydrogenation of graphite-supported platinum catalysts prepared from a zero-valent platinum complex. *Catalysis Today* 6(1-2), 171-179.

Roberts P.V., Summers R.S. (1982) Performance of granular activated carbon for total organic carbon removal. *Journal of AWWA* 74, 113-118.

Rodríguez-Reinoso F. (1998) The role of carbon materials in heterogeneous catalysis. *Carbon* 36(3), 159-175.

Ross R.A., Fairbridge C., MacCallum J.R. (1985) Carbon-fibers as supports for transition-metal catalysts in hydrocarbon oxidation reactions. *Carbon* 23(2), 209-213.

Sanga Y. (1975) Method of and apparatus for producing activated charcoal from waste tires. *US Patent* #3,875,077.

Scardi P., Antonucci P.L. (1993) XRD characterization of highly dispersed metal-catalysts on carbon support. *Journal of Materials Research* 8(8), 1829-1835.

Schafer A.I., Mauch R., Waite T.D., Fane A.G. (2002) Charge effects in the fractionation of natural organics using ultrafiltraion. *Environmental Science & Technology* 36(12), 2572-2580.

Scheible O., Gupta A., Scannel D. (1992) *Ultraviolet Disinfection Technology Assessment*, EPA 832-R-92-004. USEPA, Office of Water.

Schueller O.J.A., Pocard N.L., Huston M.E., Spontak R.J., Neenan T.X., Callstrom M.R. (1993) Crystalline nanoscale platinum(0) clusters in glassy-carbon. *Chemistry of Materials* 5(1), 11-13.

Sekine Y., Fujimoto K. (2003) Catalytic degradation of PP with an Fe/activated carbon catalyst. *Journal of Material Cycles and Waste Management* 5(2), 107-112.

Semmens M.J., Norgaard G.E., Hohenstein G., Staples A.B. (1986) Influence of pH on the removal of organics by granular activated carbon. *Journal of AWWA* 78(5), 89-93.

Shaikhutdinov S.K. (1995) Applicability of scanning-tunneling-microscopy for studying metal-catalysts supported on carbon supports. *Kinetics and Catalysis* 36(4), 549-557.

Singer P.C., Bilyk K. (2002) Enhanced coagulation using a magnetic ion exchange resin. *Water Research* 36, 4009-4022.

Singer P., Harrington G., Cowman G., Smith M., Schechter D., Harrington L. (1999) *Impacts of Ozonation on the Formation of Chlorination and Chloramination By-Products*. Denver, Colo.: AwwaRF and AWWA.

Sinha S., Amy G.L., Sohn J. (1997) Reactivity of NOM fractions in forming chlorinated DBPs. *Proceeding of AWWA Annual Conference*. Atlanta, GA.

Smisek M., Cerny S. (1970) *Active carbon*. Elsevier Publishing Company, Amsterdam.

Snoeyink V.L., Summers R.S. (1999) Chapter 13: Adsorption of Organic Compounds. In: *Water Quality and Treatment*. R.D. Letterman, ed. Mc Graw Hill Inc.

Sontheimer H., Summers R.S., Crittenden J.C. (1988) *Activated Carbon for Water Treatment*. American Water Works Association Research Foundation.

Speitel G.E., Pope, P.G., Collins, M.R., Martin-Doole, M. 2004. *Disinfection By-Product Formation and Control During Chloramination*. Denver, Colo.: AwwaRF and AWWA.

Speth T.F. (2001) Chapter 9: Activated carbon and membrane processes for disinfection by-product and microbial control. In: *Controlling Disinfection By-Products and Microbial Contaminants in Drinking Water* (EPA/600/R-01/110). Eds., Clark R.M., Boutin B.K. USEPA, Cincinnati, Ohio.

Stevens A.A., Moore L.A., Miltner R.J. (1989) Formation and control of non-trihalomethane disinfection by-products. *Journal of AWWA* 81(8), 54-60.

Stevenson F.J. (1994) *Humus Chemistry. Genesis, Composition, Reactions*. 2nd edition; Wiley Interscience: New York.

Stohr B., Boehm H.P., Schlogl R. (1991) Enhancement of the catalytic activity of activated carbons in oxidation reactions by thermal-treatment with ammonia or hydrogen-cyanide and observation of a superoxide species as a possible intermediate. *Carbon* 29(6), 707-720.

Summers R.S., Roberts P.V. (1988a) Activated carbon adsorption of humic substances I. Heterodisperse mixtures and desorption. *Journal of Colloid and Interface Science* 122, 367–381.

Summers R.S., Roberts P.V. (1988b) Activated carbon adsorption of humic substances II. Size exclusion and electrostatic interactions. *Journal of Colloid and Interface Science* 122, 382–397.

Summers R.S., Hooper S.M., Solarik G., Owen D.M., Hong S. (1995) Bench-scale evaluation of GAC for NOM control. *Journal of AWWA* 87(8), 69-80.

Summers R.S., Hooper S.M., Shukairy H.M., Solarik G., Owen D. (1996) Assessing DBP yield: Uniform formation conditions. *Journal of AWWA* 88(6), 80-93.

Suzuki Z. (1954) Chemical studies on active carbon. II. Activation process. *Nippon Kagaku Zasshi* 75, 389-392.

Symons J.M., Krasner S.W., Sclimenti M.J., Simms L.A., Sorensen H.W. Jr., Spietel G.E. Jr., Diehl A.C. (1996) Influence of bromide ion on trihalomethane and haloacetic acid formation, In: *Disinfection By-products in Water Treatment: The Chemistry of Their Formation and Control*. Eds., Minear R.A., Amy G.L. Lewis Publishers, Florida.

Tamon H., Okazaki M. (1996) Influence of acidic surface oxides of activated carbon on gas adsorption characteristics. *Carbon* 35(6), 741-746.

Teng H., Wang S.C. (2000) Preparation of porous carbon from phenol-formaldehyde resins with chemical and physical activation. *Carbon* 38, 817-824.

Thurman E.M., Malcolm R.L. (1981) Preparative isolation of aquatic humic substances. *Environmental Science & Technology* 15(4), 463-466.

Thurman E.M., Wershaw R.L., Malcolm R.L., Pinckney D.J. (1982) Molecular size of humic substances. *Organic Geochemistry* 4, 27-35.
Thurman E.M. (1985) *Organic Geochemistry Of Natural Waters*. Martinus Nijhoff/Dr. W. Junk Publishers, Dordrecht/Boston/Lanscaster.
Umehara T., Harriott P., Smith J.M. (1983) Regeneration of activated carbon. *AIChE Journal* 29, 732-741.
USEPA (1996) ICR Manual for Bench- and Pilot-Scale Treatment Studies. EPA 814-B-96-003.
USEPA (1998) National Primary Drinking Water Regulations: Disinfectants and Disinfection By-products; Final Rule. *Federal Register* 63(241): 69390-69476, Dec. 16, 1998.
USEPA (2001) United States Environmental Protection Agency document EPA/816/R-01/014, June 2001.
USEPA (2006) National Primary Drinking Water Regulations: Stage 2 Disinfectants and Disinfection Byproducts Rule; Final Rule. *Federal Register* 71 (2): 388-486. Jan. 4, 2006.
National Primary Drinking Water Regulations: Stage 2 Disinfectants and Disinfection Byproducts Rule; Final Rule. Dec. 15, 2005.
Vinke P., van Der Eijk M., Verbree M., Voskamp A.F., van Bekkum H. (1994) Modification of the surfaces of a gas-activated carbon and a chemically activated carbon with nitric-acid, hypochlorite, and ammonia. *Carbon* 32(4), 675-686.
von Blucher H., de Ruiter E. (1999) Process for producing granulated activated carbon. *US Patent* #5,977,016.
Vuorio E., Vahala R., Rintala J., Laukkanen R. (1998) The evaluation of drinking water treatment performed with HPSEC. *Environment International* 24(5/6), 617-623.
Warta C.L., Papadimas S.P., Sorial G.A., Suidan M.T., Speth T.F. (1995) The effect of molecular oxygen on the activated carbon adsorption of natural organic matter in Ohio River water. *Water Research* 29, 551-562.
Weber W.J., Jr., Pirbazari M., Long J.B., Barton D.A. (1980) Potential mechanisms for removal of toxic and/or carcinogenic compounds from water supplies. In: *Activated Carbon Adsorption of Organics from the Aqueous Phase.* Vol. I. Eds., Suffet I.H., McGuire M.J. Ann Arbor Science, Ann Arbor, Michigan.
Weber W.J., Jr., Voice T.C., Jodellah A.M. (1983) Adsorption of humic substances: the effects of heterogeneity and system characteristics. *Journal of AWWA* 75, 612–618.
White M.C., Thompson J.D., Harrington G.W., Singer P.C. (1997) Evaluating criteria for enhanced coagulation compliance. *Journal of AWWA* 89(5), 64-77.
Whittaker M.P., Grindstaff L.I. (1974) Production of activated carbon from rubber and a carbonaceous binder. *US Patent* #3,822,218.
Wolfe R., Ward N., Olson B. (1984) Inorganic chloramines as drinking water disinfectants: A review. *Journal of AWWA* 76(5), 74.
Wu F.C., Tseng R.L., Hu C.C. (2005a) Comparison of pore properties and adsorption performance of KOH-activated and steam-activated carbons. *Microporous and Mesoporus Materials* 80, 95-106.
Wu F.C., Tseng R.L., Juang R.S. (2005b) Comparison of porous and adsorption properties of carbons activated by steam and KOH. *Journal of Colloid and Interface Science* 283, 49-56.
Wu F.C., Tseng R.L., Hu C.C., Wang C.C. (2005c) Effects of pore structure and electrolyte on the capacitive characteristics of steam- and KOH-activated carbons for supercapacitors. *Journal of Power Sources* 144, 302-309.

Yang C.M. El-Merraoui M. Seki H., Kaneko K. (2001) Characterization of nitrogen-alloyed activated carbon fiber. *Langmuir* 17, 675-680.

Yang C.M., Kaneko K. (2001) Adsorption properties of nitrogen-alloyed activated carbon fiber. *Carbon* 39, 1075-1082.

Young C.T. (1996) Nuts. *Kirk-Othmer Encyclopedia of Chemical Technology* 17, 579.

Zaidi A.H. (1987) Stability of Pd-C phase in the oxidation of ethylene over palladium catalysts. *Applied Catalysis* 30(1), 131-140.

Zhou Q., Cabaniss S.E., Maurice P.A. (2000) Considerations in the use of high-pressure size exclusion chromatography (HPSEC) for determining molecular weights of aquatic humic substance. *Water Research* 34(14), 3505-3514.

ABBREVIATIONS

^{13}C-NMR	^{13}C-Nuclear magnetic resonance
ACF	activated carbon fiber
AWWA	American Water Works Association
AwwaRF	Awwa Research Foundation
BET	Brunauer-Emmett-Teller equation
C/F/S	coagulation/flocculation/sedimentation processes
d	the diameter of carbon particle
D/DBPR	Disinfection/Disinfection By-Products Rule
DBP	disinfection by-product
DDW	distilled deionized water
DFT	density functional theory
TOC	dissolved organic carbon
DOM	dissolved organic matter
DR	Dubinin-Radushkevich equation
EBCT	Empty bed contact time
$EBCT_{LC}$	Empty bed contact time of large column simulated by RSSCT
$EBCT_{SC}$	Empty bed contact time of small column (used in RSSCT)
GAC	granular activated carbon
GC/MS	Gas chromatograph(y)/mass spectroscopy
HAA	haloacetic acid
HAA_9	nine haloacetic acid species
HCl	hydrochloric acid
HD	HD3000 carbon
HPO	hydrophobic DOM fraction
HPSEC	high pressure size exclusion chromatography
ICP-AES	inductively coupled plasma atomic emission spectroscopy
IR/FTIR	infrared radiation/Fourier transform infrared absorption spectroscopy
IS	ionic strength
IUPAC	International Union of Pure and Applied Chemistry
L	the length of the carbon bed
LC (Subscript)	large column simulated by RSSCT
m	mass of carbon used for packing one column
MB	Myrtle Beach
MBR	Myrtle Beach raw water
MBT	Myrtle Beach treated water
MCL	maximum contaminant level
MIEX	magnetic anion exchange resin
MW	molecular weight
MW_w	weight average molecular weight
NOM	natural organic matter
pH_{PZC}	pH of point of zero charge
Q	flow rate of influent
R	raw water

RO	reverse osmosis
RSD	relative standard deviation
RSSCT	rapid small scale column test
SAC	steam-treated carbon
SC (Subscript)	small column used in RSSCT
SEC	size exclusion chromatography
SF	Scaling factor (the ratio of the carbon particle size used in large scale experiment and RSSCT)
SOC	synthetic organic contaminant
SP	Spartanburg
SpR	Spartanburg raw water
SpT	Spartanburg treated water
SUVA	specific ultraviolet absorbance
$SUVA_{254}$	specific ultraviolet absorbance at 254 nm
T	treated water
THM	trihalomethane
TOC	total organic carbon
UFC	uniform formation condition
USEPA	United States Environmental Protection Agency
UV	ultraviolet absorbance
UV_{254}	ultraviolet absorbance at 254nm
v_{sc}	hydraulic loading (superficial velocity)
WTP	Water treatment plant
XPS	X-ray photoelectron spectroscopy
ε	the porosity of carbon bed
ν	kinematic viscosity of water
ρ_b	carbon bed density

IWA Publishing's authorised EU representative for General Product Safety Regulations is Diane D'Arras, 15 rue Duret, 75116 Paris, France, e-mail: safety@iwap.co.uk.

Printed and bound by CPI Group (UK) Ltd, Croydon, CR0 4YY
17/09/2026
02218808-0001